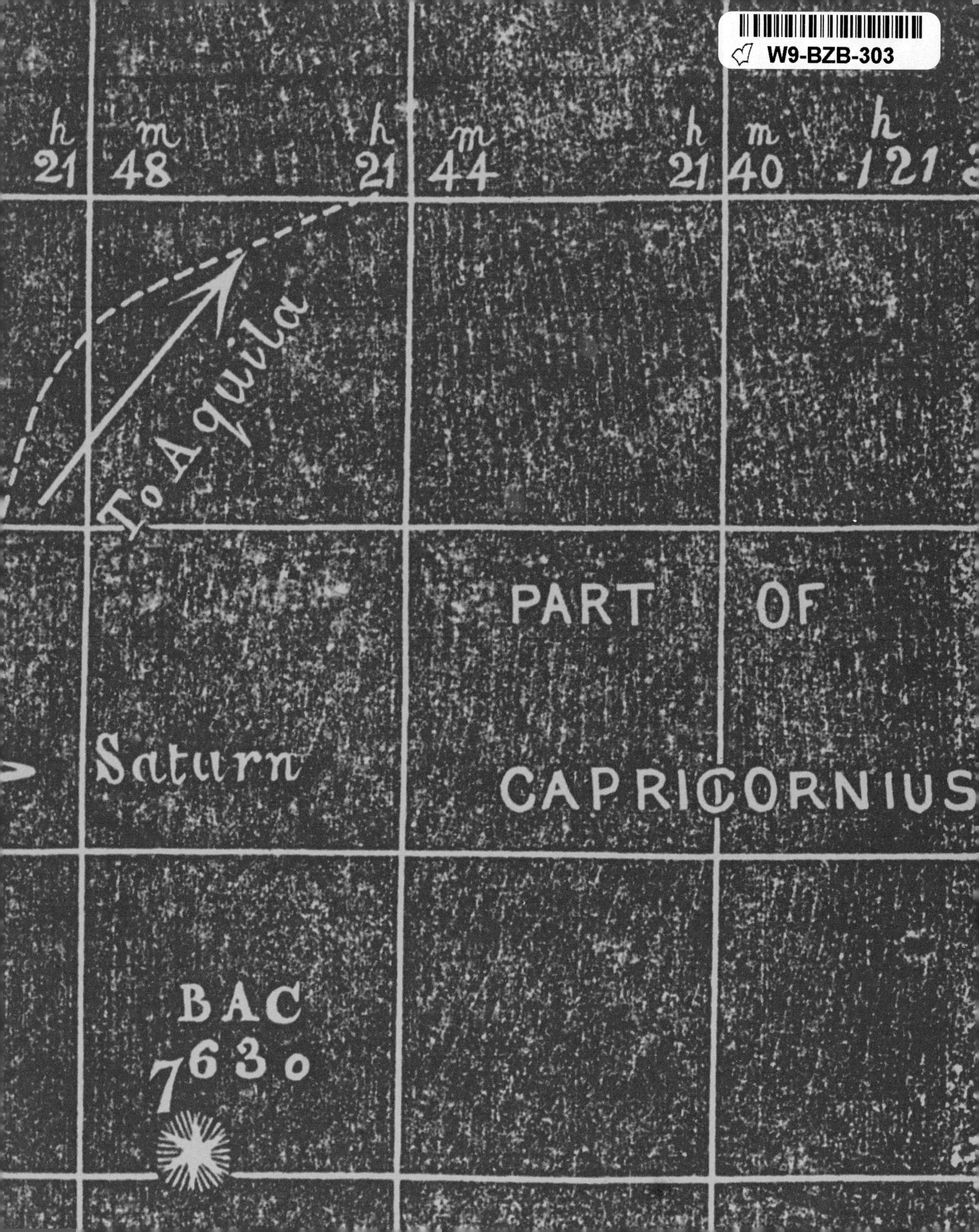

W9-BZB-303
h m
21 48
h m
21 44
h m
21 40
To Aquila
Saturn
PART OF
CAPRICORNIUS
BAC
7630

THE UNIVERSE

AN ILLUSTRATED HISTORY OF ASTRONOMY

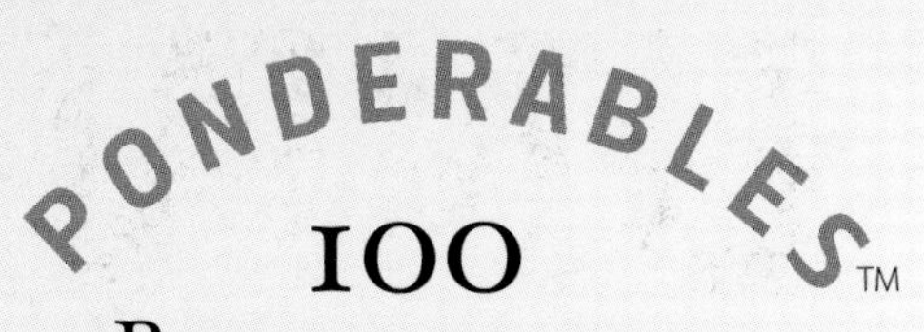

Breakthroughs That Changed History

WHO DID WHAT WHEN

THE UNIVERSE

AN ILLUSTRATED HISTORY OF ASTRONOMY

Tom Jackson

SHELTER HARBOR PRESS
NEW YORK

Contents

Title page: An image of the sky taken by the Wide-field Infrared Survey Explorer (WISE) orbiter shows variations in the infrared radiation, or heat, coming from deep space. The Milky Way, our own galaxy, forms the blue disc in the middle of the image.

REACHING FOR THE STARS

THE NEXT FRONTIER

Introduction

ASTRONOMY BEGINS WITH SOME BIG QUESTIONS: WHERE AM I, AND WHERE DID I COME FROM? As thinkers of all types pondered their existence, the stars frequently featured in the answers, perhaps as geomantic objects showing the way forward, navigational aids (really showing the way), or as steady, constant points against which to measure the Universe.

Something like the smartphone of the first millenium, the astrolabe was watch, map, fortune teller, and compass all in one.

A thousand-year-old rock painting shows that, like everyone else, Native Americans in New Mexico were following events in the sky.

To study the heavens takes a lot of imagination. Our understanding of the Universe has taken shape within the abstracted minds of countless philosophers, sages, and scientists. And that is where it largely remains. We can't visit other stars to take a look, and less than 500 of us have had the privilege to slip the shackles of gravity and look down on Earth from space. Our best view of neighboring planets has always been through a lens. Instead the structure of the Universe took shape in our minds, passing down from astrologers to navigators and then to scientists.

Every step along the path of that knowledge, along the stream of astronomical ideas, is a story in itself, and here we present the hundred best all together. Each story relates a ponderable, a weighty problem that became a discovery and changed the way we understood the Earth, the stars, the whole Universe—and our place in it.

A PONDERABLE

The search for knowledge is an unending work, with understanding hewn from a bedrock of evidence, following intuitions that become theories, before being solidified into fact. Every new ponderable provides another detail, a finishing touch or powerful reshaping, of our world view—where we fit, who we are... and whether we are alone.

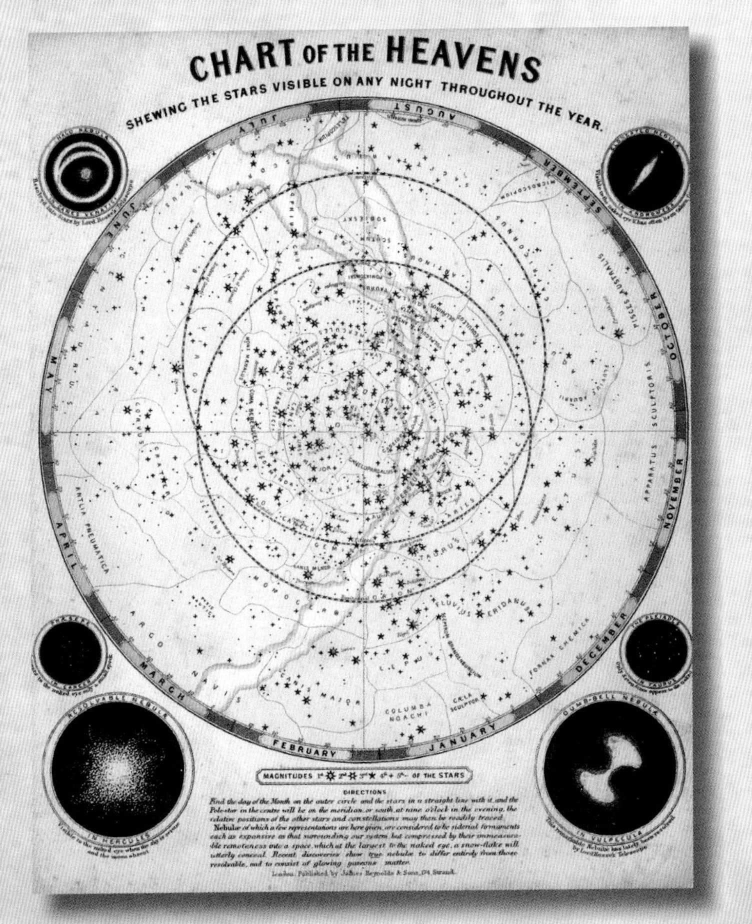

The industrial advances of the 19th century made complex star maps and simple telescopes widely available and turned astronomy into an amateur affair. Even today, many celestial objects are first revealed in the observations of hobbyist stargazers.

A breathtaking view of the stars on a clear night is reason enough to explain why early astronomy was infused with magic and divinity. Perhaps the first star catalogs were attempts to understand the gods better and so predict what the future held. However, another human attribute, a love of patterns, was soon a driving force, and astronomers from Mexico to China began to codify the heavenly orchestration of lights across the night's sky.

EXCEPTIONS SHOW THE WAY

Objects that did not follow the same rules would have stood out among the carefully recorded data. And it was these exceptional bodies—the planets, comets, nova (or newly appeared stars) and cloudy swirls among the points of starlight—which provided the first clues that helped to solve the many mysteries of the cosmos.

We have a very detailed history of the Universe now, or at least we think we do. There is an inconceivable vastness in which to discover more anomalies that might change the accepted version of events—not for the first time. Modern astronomy has diversified like other sciences to encompass astroseismologists, who monitor the quaking insides of stars, astrobiologists who search for places where life might form outposts, and cosmologists who consider the big picture. As it stands astronomers can still only see one per cent of that picture, all else is dark—quite literally. Will we ever see it all?

Modern astronomers look further into space and deeper back in time, but also revisit the old sights with new technologies. This image shows the ultraviolet light (invisible to the naked eye) released during a severe solar storm. The bubble-like plumes, or mass ejections, more than double the size of the Sun in a matter of hours.

Scale of the Universe

THE UNIVERSE IS CLEARLY ENORMOUS, BUT EXACTLY HOW LARGE IT IS DEFIES OUR IMAGINATION. We can approach an understanding of its size when the measurements are written down or shown in pictures, but against all that emptiness—pricked only occasionally by truly gargantuan objects—the human scale, our sense of personal space and of our place within the Universe, seems dwarfed, shrinking almost to nothingness.

This diagram shows the relative sizes of the Solar System's planets and their moons but not the distances between them. Nor is the Sun to scale.

Sun

Mercury Venus Earth Mars Asteroids Jupiter Saturn Uranus Neptune

GETTING THE MEASURE OF ASTRONOMY

Eratosthenes, the first person to measure the size of Earth using objective evidence, found it to be 252,000 stadia. This unit was one length of the local arena that Alexandrian athletes dashed up and down, perhaps wearing full armor—or nothing all. The mile, another very old measure, albeit one still in use, was the distance traveled in 1,000 steps by a marching Roman legion. The metric meter, upon which all scientific measurements are now based, was initially defined as a ten-millionth of the distance from pole to equator. These units all work on the human scale and are still meaningful for Earth-spanning distances, but they soon become unwieldy on the astronomical scale: it is 42 billion meters to Venus (on a good day) and 356 million meters to the Moon. Even to our nearest neighbors, the numbers are too large to imagine.

INSIDE THE SOLAR SYSTEM

When pacing out the Solar System, our patch of the Universe, astronomers use the astronomical unit, AU. One AU is the average distance from the Earth to the Sun. That's about 150 million km (93 million miles). The AU is very convenient. How far are we from the Sun? One AU. At its closest Earth comes within 0.3 AU of Venus, 0.5 AU of Mars, and the orbit of Neptune is 30 AU out. However, that is just the beginning. The Solar System stretches at least five thousand times further in all directions. Now the AU is becoming less useful, and the next nearest star is 268,305.24 AU from the Sun. We need a new unit.

EVERYWHERE ELSE

Just about all that we know from outside the Solar System arrives in the form of light or another form of electromagnetic radiation (radio waves, X rays, and the like). All of this moves at the same speed—a little over 7 AU per hour, or 299,792,458 meters per second. The light from the nearest star, Proxima Centauri, takes 4.24 years to reach us, and so it is 4.25 light-years away. Voilà, a new unit. A light-year is about 63,000 AU (10 trillion km or so). The visible Universe extends 13.7 billion light-years in all directions. Perhaps we'll need yet another unit one day.

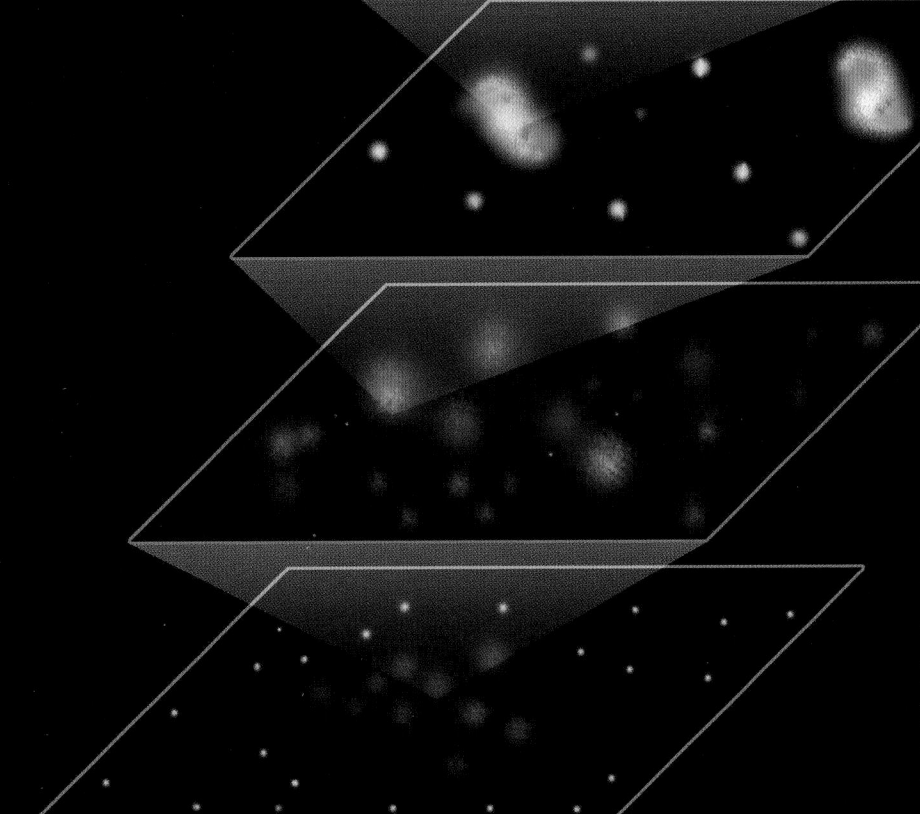

The Solar System
The Earth is eight light-minutes from the Sun. (That is how long sunlight takes to reach us.) Jupiter is 40 light-minutes away while Neptune is four light-hours out. The Solar System peters out by about 0.5 light-years.

The Nearby Stars
The nearest star, Proxima Centauri, a dim red dwarf lurking just this side of the Alpha Centauri system, is 4.24 light-years away. The next 15 nearest stars are within 11 light-years of the Sun.

The Galaxy
Our Solar System is in the Orion Arm of the Milky Way. This arm is 3,500 light-years across. The entire galaxy is 100,000 light-years wide.

The Local Group
The Milky Way is the second largest galaxy in the Local Group, after Andromeda. This cluster has 50-odd more galaxies which fill a space 10 megalight-years across.

Virgo supercluster
The Local Group is one of dozens of other galaxy clusters (more than 100 in total) in a supercluster that is 110 mega-light years across.

The Observable Universe
There are millions of superclusters in the Universe, often forming filaments or "great walls," half a billion light-years long. As far as we know, the Universe is 13.7 billion years old, since light from further away than that has not reached us yet. So at the moment, 13.7 light-years is the furthest that we can see. Perhaps the Universe is bigger... but so far its light hasn't reached us.

AT THE CENTER OF THE UNIVERSE

1 Monuments to the Stars

The megaliths of Stonehenge are perhaps the most iconic prehistoric monuments (thanks in part to being concreted safely back into place over the last century). The debate over the standing stones' true function continues, ranging from an acoustic arena to a healing center. Its most verifiable feature is as a solar calendar, with the midsummer sunrise framed by its arches.

ASTRONOMY IS AS OLD AS HUMANITY ITSELF. Our prehistoric ancestors made out patterns in the starlight that filled the dark of ancient nights. It appears that many of the structures that remain from those days were constructed to reflect the awesome and unending motion of the heavens.

The human mind is built to find patterns, to make out the outline of a predator lying in ambush, to map food and water sources, and to think through alliances with friends and strangers. It does not require a large leap of imagination to see how, over the generations, early human cultures made a strong connection between the rhythm of seasons and the periodic appearance of heavenly bodies. Astronomy was born.

SHADOW SNAKE

El Castillo (below), the main pyramid at the Mayan complex of Chichen Itza in Mexico, has 365 steps, one for each day of the year. The pyramid is a temple to Kukulkan, a flying serpent god. At the foot of the steps on the northern side is a carved snake's head and on the spring and fall equinoxes, the sun casts a snake-shaped shadow on the staircase, showing Kukulkan slithering out of the sky.

The link between the stars and seasons became of crucial importance with the rise of agriculture. Sowing seed too early or late could spell a lingering death from hunger. The stakes could not be higher, and primitive cultures steeped in superstition would do what they could to insure the heavenly forces were favorable to them. This explains why so many early civilizations poured millions of man-hours into erecting stone monuments to the gods of the sky, many of which were constructed well enough to survive to this day. Some, like Stonehenge, captured the sun at turning points such as an equinox (day and night of equal length) or solstice (longest or shortest days of the year). Others reflected the heavens to insure a good connection with the gods. The square-based Pyramids of Giza in Egypt are orientated to the points of the compass. However, since the compass was not invented until 2,500 years after their completion, surveyors must have used stellar alignments with the Pole Star to get the megatombs into their auspicious positions.

2 Following the Sun and Moon

SYSTEMATIC RECORDS OF THE MOTION OF THE SUN, MOON, AND CERTAIN BRIGHT STARS formed the basis of the first calendars. Then early astronomers took it further, and used the data to predict celestial events, such as eclipses.

A Babylonian tablet records the sighting of a comet in 163 BC. Later analysis revealed that this was the same body we call Halley's Comet.

By 2000 BC, Egyptian and Babylonian astronomers had established the approximate length of the year as 365 days. They were unaware that this period was the length of time it took Earth to travel once around the Sun. Instead, the Egyptians based their year on the appearance of Sirius, the Dog Star, which coincided with the annual Nile flood.

The other basic time units, the day and month, were also based on astronomical events—the rising and setting of the Sun, and the phases of the Moon respectively. The Chinese, Babylonians, and perhaps astronomers from other cultures could track the positions of the Sun and Moon accurately enough to predict eclipses. Thales of Miletus, a seminal figure in science, predicted a solar eclipse in 585 BC. Legend has it that this event led to the end of a long war between the Greeks and Persians.

3 Finding Patterns

IT IS PERHAPS ONLY NATURAL THAT HUMANS SHOULD PROJECT THEIR MYTHS, THEIR STORIES OF DIVINE CREATION AND SUPERHUMAN EVENTS, onto a backdrop of stars that was literally out of this world.

Constellations, the imaginary patterns in the night's sky, reflect the imagery of a culture. We are all more or less familiar with the dogs, bears, hunters, and heros that dominate the ancient Greek constellations and which form the basis for how modern astronomers divide up the sky today. These constellations are used to officially describe regions of the sky. Other star patterns are "unofficial."

Same stars, several stories

Perhaps the best way to see the strong links between constellations and the cultures that create them is to look at one of the most famous constellations. What the Romans Latinized as Ursa Major was originally a large (or great) bear that appeared to the Greeks. (This was the bigger of two that filled neighboring areas of sky. According to myth, they were a mother and son caught up in a jealous spat between Zeus and his wife.) However, to later eyes the seven brightest stars in the Great Bear became the Plough, and later still in North America, the Big Dipper, resembling a giant ladle.

In Hindu culture the Big Dipper is known as the Seven Sages, named for some important figures in Vedic literature. The seven stars also get a mention in the biblical Book of Amos, while the same grouping has been identified in rock carvings at a grave site in Puyang in the Chinese province of Henan, dating from 4,000 BC.

The Milky Way is formed by the combined light of billions of stars clustered together into our galaxy.

THE MILKY WAY

In East Asia it is the Silver River; in India it is the Ganges of the Heavens; in the Middle East and Africa it is better known as the Straw Road, while in Central Asia it is the Bird's Path. All these refer to the pale strip of light that runs across the night's sky, which is officially termed with its European name, the Milky Way. This celestial feature is often hard to see with modern light pollution and even bright moonlight drowns it out. However, in the right conditions, what the Romans identified as *via lactea* makes a glorious sight. The Latin term came from the Greek *galaktikos kylos*, meaning "milky circle." So what are we looking at? It is a view into our galaxy—still named the Milky Way— from where our Sun is located. The word *galaxy* is also derived from the Greek for milk.

Modern shapes

Strictly speaking the Big Dipper and its related forms is not a constellation, but an asterism—an unofficial pattern of stars. As well as being easy to recognize and always visible in northern skies, this asterism is also well known because the two stars farthest from the handle are the Pointers, tracing a line that takes the amateur astronomer, boy scout, or beleaguered sailor straight to the North Star, which as its name suggests is always to the north.

The Summer Triangle is another asterism, first described in the 1920s as a relatively empty area of sky between the constellations of Aquila (the eagle), Lyra (the harp), and Cygnus (the swan) but popularized in the 1950s by British television astronomer Patrick Moore, who suggested it as something for his stargazing audience to look at in the summer months, when most other features of the northern sky were harder to spot.

The Dresden Codex, named after the German city in which it is housed, is a work of Mayan literature, written on sheets of bark like this one. It is about 800 years old but its contents are thought to hail from a few centuries before. Many of its 78 pages are concerned with astronomical data including the World Tree, a constellation based on the Milky Way.

This drawing shows the constellation of Taurus the Bull in a 15th-century copy of The Book of Fixed Stars, *by 10th-century Arab astronomer Al Sufi. The book consolidates the Greek constellations with the astronomical traditions of Arab scientists.*

History of constellations

The set of Greek constellations we use today were probably consolidated in the 4th century BC. Obviously—and not without irony—these patterns did not appear out of thin air. Many of them relate myths of Mycenaean origin (around 1000 BC), mostly ending with Zeus, the king of gods elevating the characters to the heavens to honor them or save them from an earthly torment of some kind or other. Orion the Hunter is another familiar constellation and has a most engaging place in the story told among the stars. He can be seen with his hunting dogs (Canis Major and Canis Minor) eyeing up a bull (Taurus) and a hare (Lepus) as possible quarry. One story relates that Orion refused to give a goddess his bow, and a thief sent to steal it ended up killing the hunter by accident. This is why Orion disappears completely below the horizon in spring. Another story has Orion hunting with the goddess Artemis, much to the displeasure of Apollo, her brother, who kills Orion with a scorpion sting. As the constellation Orion sets in the west each night, Scorpio (the killer scorpion) rises in the east, locked in an eternal chase across the sky.

The Greek constellations do not cover the whole sky, much of the Southern Hemisphere is blank—it was not visible from the classical Greek world. The position of this blank region leads astronomers to suggest that the modern constellations date from 1130 BC as viewed from around 33° North—the approximate latitude of the Mesopotamian civilizations from that era.

4 Fixed and Wandering Stars

THE WORD *ZODIAC* IS PERHAPS MORE ASSOCIATED WITH ASTROLOGY—THE HIGHLY DOUBTFUL PRACTICE OF PREDICTING THE FUTURE BY THE POSITION OF THE PLANETS. However, the term, which means circle of animals in Greek, has a thoroughly astronomical basis.

Many of the terms and concepts that we apply to modern astronomy were passed down to us by ancient Greek astronomers. In all likelihood they were reflecting earlier ideas formed in Babylon or even further from Greece. Eudoxus, a 4th-century associate of Plato, is credited as the best source on the state of classical astronomy. His compilation of constellations is the one we still use for the Northern Hemisphere. Of course, early astronomers in China, India, and other countries used different constellations.

This clay disc made in Alexandria probably in the 1st century BC, shows the 12 signs of the zodiac, most of which are still in use today.

Moving objects

Eudoxus also included the Babylonian concept of the zodiac in his description of the stars. This was a band of sky which was populated by some very unusual objects—the wanderers. The Greek word for wanderer has a modern translation: planet. However, in the crudest terms, the wanderers were the Sun, Moon, and five smaller stars (which we now understand to be planets).

The largest and brightest of them—the Sun—traces a path through the sky known as the ecliptic. The name relates to how an eclipse occurs when the Moon is near (or very near) this line.

Staying close

The Moon and planets never strayed far from the ecliptic (less than 10° either side)—the aforementioned zodiac. The twelve constellations in the zodiac and the path of the wanderers within them took on special significance for soothsayers and natural philosophers alike. The former attempted to predict the future by relating birth dates to the paths of the seven wanderers. To philosophers, the zodiac's prime movers were the pieces of a puzzle that could reveal Earth's place in the Universe.

5 Gods in Space

THE WANDERING STARS, FREE SPIRITS AMONG THE RIGID FRAME OF CONSTELLATIONS, WERE ASSOCIATED WITH GODS, each with a particular personality. Scientific convention has adopted the Roman names for the planets, and their supernatural qualities linger to this day.

Eudoxus would have known the teachings of mathematician Philolaus on the motion of stars. Philolaus placed Earth in motion around a central fire, with the Sun, Moon, and planets moving in widening circles around this same point. Adding an outer sphere upon which the constellations were fixed resulted in nine heavenly objects in perpetual motion. As a Pythogorean, Philolaus saw nine as an imperfect number, but ten was flawless. So he proposed that an invisible object was situated between the fire and Earth moving in an eternal eclipse, shielding the primordial flames from view.

The Romans identified the brightest and most frequently seen planet with Venus, the goddess of love and fertility. She balanced the influence of Mars, with whom she had a son, called Cupid, the god of desire.

Deities at work

The wandering Moon, Sun, and five planets, therefore, were gods going about their business, watching over and influencing life on Earth. The characteristics of the planets became one with the deity they represented. Mercury (Hermes in Greek) was the adolescent messenger of the gods, always moving quickly, frequently disappearing from view only to reappear as suddenly, and generally an unreliable presence. We know the next planet as Venus, the Roman goddess of love. However, to the Greeks this was actually twinned stars: Hesperus was the evening star, the first one appearing at sunset; Phosphorus rose in the morning, and was later identified as Lucifer (the Shining One)—an altogether different story.

The science of astronomy and superstitions of astrology were not clearly separated until the 19th century, and observations of planets and other heavenly bodies were often interpreted as good or bad omens.

The angry red of the next planet made it Mars (Ares in Greek), the god of war and farming. His month, March, signaled the best time to start a war or sow the fields (but not both). Jupiter (Jove or Zeus), a slow-moving but stately presence, was the king of the gods. In many traditions, the chief god dethrones his father. Jupiter did just that to his, Saturn, (Cronos, the god of time, in Greek). Saturn was left to wander just beneath the fixed stars. Was he the last one? After all, Saturn usurped his own father Caelus, better known as Uranus. Perhaps he was out there too. Time would tell.

6 Centering Earth

ONE OF THE TENETS OF MODERN ASTRONOMY IS THAT THE LAWS OF PHYSICS AS OBSERVED ON EARTH ALSO HOLD AT ALL POINTS IN THE UNIVERSE. Aristotle applied this notion to his orderly Universe, removing Philolaus' central fire and placing Earth at the heart of everything.

While Aristotle's model of the Universe was frequently questioned by scientists, criticisms of it were censored first by a Greek belief in the perfect harmony of nature and then later by the Catholic Church which upheld Aristotle's views as orthodox religious teaching. As a result the Aristotelian Universe was accepted fact from the fourth century BC to the 1600s.

Layers of matter

Aristotle began with the idea that universe was constructed of four elements: earth, water, air, and fire. These earthly materials combined to form all natural substances. Heat, aridity, cold, and damp were evidence of the presence of a corresponding element. The smoke from smoldering wood was the air escaping from within, the resin forced out by the heat was the water, the ash left behind was the earthen component, while the flickering flames were its fire.

Aristotle reasoned that the driving force behind nature was the desire of elements to separate into their pure forms. Earth was the most basic of elements and the heaviest, so it sunk beneath the rest, creating land. Water formed the next layer, followed by air and then fire. Volcanic eruptions, earthquakes, and rain were all evidence of the elements finding their ways to their rightful positions.

The ring of fire reached as far as the Moon, while beyond lay the Sun and the five planets, all circling Earth. Everything was enclosed by a celestial sphere where the fixed stars were embedded in crystal.

The heavenly region of the Universe beyond the Moon was filled with ether. This, according to Aristotle, was the *quintessence,* a fifth, heavenly element located beyond the reach of humans, never mixing with the more lowly elements. Even after the true nature of the Solar System was uncovered, the concept of an all-pervading ether continued into the twentieth century, only disproved by Einstein's theory of special relativity.

Published in 1539, a few decades before being debunked by Nicolaus Copernicus, Peter Apian's Cosmographia *contains a map of the Universe as perceived by Aristotle complete with zodiac.*

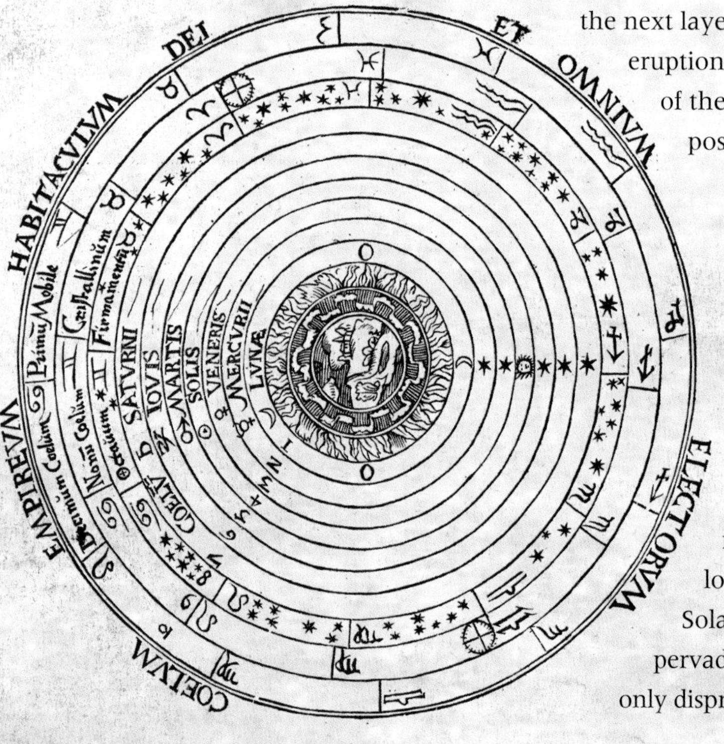

7 A Turning Sphere

IT IS EASY TO LABEL ARISTOTLE AS THE ONE WHO GOT IT ALL WRONG. HOWEVER, HIS PHILOSOPHY DID CONTAIN SOME TRUTH. The most significant in astronomical terms was his insistence that Earth was a sphere, and his evidence for this came from elephants.

Most ancient cultures appear to have settled on the fact that the world was some kind of rounded shape. In the seventh century BC, the Greeks thought they lived on a disk, although in 580 BC Anaximander of Miletus suggested it was more of a cylinder, with a flat landmass on top and a seething ocean curving around it.

One of Anaximander's pupils (according to some sources), was Pythagoras, the famous mathematician who possibly assimilated ideas from Egypt, Mesopotamia, and perhaps further afield. He proposed that all heavenly bodies were spheres, and as Earth was one of them, well, it must be a sphere too. Not everyone agreed. Democritus, the chief exponent of atomism (the idea that the Universe was made of tiny units called atoms) is much hailed as a visionary genius whose ideas were quashed by blind allegiance to Aristotle. However, Democritus was also fallible: he supported the flat Earth concept.

Aristotle's work tells us that ancient astronomers understood lunar eclipses as the shadow of Earth passing over the Moon. Curiously, the red of a total lunar eclipse is the result of sunlight scattered through Earth's atmosphere.

Aristotle's logic

Pythagoras never set out any reasoning but Aristotle did better 200 years later: the Moon was illuminated by light from the Sun (as was Earth). That meant that the phases of the Moon were not the result of the Moon changing shape, but a changing view of the half-illuminated body from Earth. As that view changed, the shape of the terminator—the boundary between shadow and light—was always curved. The only shape that behaved like this was a sphere. During a lunar eclipse, the shadow of Earth edging across the Moon was also always rounded. Only round shapes always cast round shadows. But why was Earth not a flat disk? Aristotle pointed out that traveling on journeys to the far north or south meant that the angles between the stars and horizon changed. For example, the North Star became lower in the sky as you went south. This is only possible if traveling on a curved surface. Finally Aristotle played the elephant card: elephants live in India (the eastern limit of the then known world) and Morocco (the western limit) and so the surface of world must be interconnected!

8 A Heliocentric Theory

A LITTLE MORE THAN A GENERATION AFTER ARISTOTLE, A GREEK CALLED ARISTARCHUS OF SAMOS PROPOSED AN ALTERNATIVE TO THE GEOCENTRIC (Earth-centered) view of the Universe. His theory put the Sun in the middle.

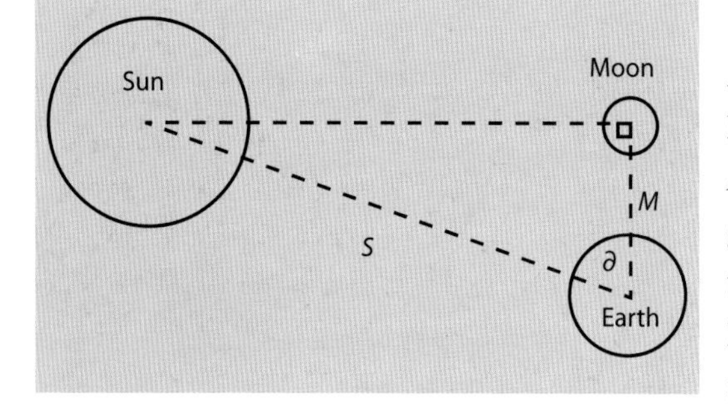

Aristarchus used trigonometric functions to calculate the ratio of M and S from the angle ∂. In reality, distance S is far greater than M and so angle ∂ is much closer to 90° than shown here.

Aristarchus's *heliocentric* idea comes down to us second hand. His only surviving text *On the Sizes and Distances of the Sun and Moon* (published around 250 BC) makes no mention of it but gives some hints. In it Aristarchus uses trigonometry to show that the Sun was about 19 times farther away than the Moon. He postulated that during a half moon, the Moon, Earth, and Sun form a right-angled triangle. He then says that the angle of the Sun is 87° (3° off the right angle). There is no record of how or if he measured this—and modern astronomers would have difficulty doing so. It was wrong anyway, so it is likely that Aristarchus just estimated it for a bit of geometrical gymnastics: the Sun looks to be at 90° but Euclid's law of triangles tells it cannot be. In any event, Aristarchus continued his exercise by using his distances to estimate the size of the Sun, Moon and then the Earth from the size of shadow cones in a lunar eclipse. Again he was wildly off, finding that the Sun was seven times the size of Earth (it is 109 times the size). It is pure speculation, but perhaps it was finding that the Sun was the largest object in the sky that led to Aristarchus making it the central point. We can only speculate again what history would have been like if Aristarchus's ideas had been taken more seriously.

9 Eratosthenes Measures Earth

AT THE END OF THE 3RD CENTURY BC, ANOTHER MATHEMATICIAN DEVISED A REMARKABLY SIMPLE WAY OF CALCULATING THE SIZE OF EARTH that required him to take just one single measurement.

As Aristotle had pointed out in his treatise on the shape of Earth, it was widely known that heavenly objects reached different altitudes—measured as angles above the horizon—in different areas of the ancient world. The Sun was no exception. When Eratosthenes, the chief librarian of Alexandria, heard that although the Sun cast shadows in Alexandria at noon on the first day of summer, it did not do so in Syene, a

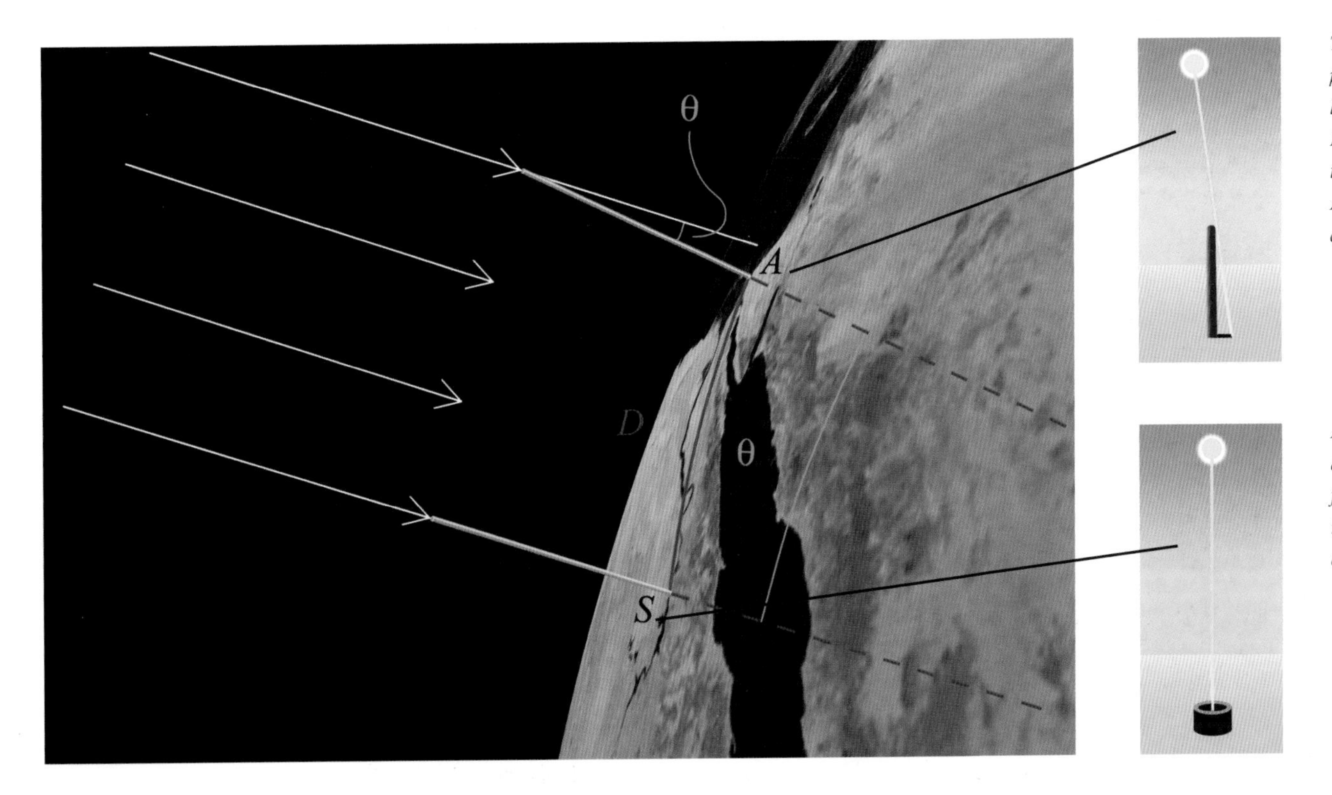

The triangle formed by pillar, shadow and light beam was enough for Eratosthenes to calculate the angle of the Sun above Alexandria on the first day of summer.

At Syene, the Sun was directly overhead on the first day of summer and illuminated the bottom of a well.

city to the south now known as Aswan, he was intrigued. The evidence from Syene came as a report that the noon-day Sun illuminated a well on Elephantine Island in the Nile near Syene. Eratosthenes realized that if the Sun was demonstrably overhead Syene on a specific day (an altitude of 90°), he could measure the lower altitude over Alexandria at the same moment, and use that figure to calculate the circumference of Earth.

The method and result

Eratosthenes reasoned that the light from the Sun traveled in parallel beams. The beams over Syene arrived vertically, while they approached Alexandria at an angle—casting shadows. He already knew, from the archives of his library and by consulting traveling merchants, that the distance between the two cities was 5,000 stadia (an ancient measurement, shown as distance *D* above). All he had to do was calculate the angle of the sunlight in Alexandria (marked as theta). He did this by measuring the length of a shadow cast by a gnomon (a vertical stake) at noon on the summer solstice—geometry did the rest.

The angle of the light measured from the gnomon in Alexandria was the same as the angle between the two cities as measured from the center of the Earth. The figure Eratosthenes arrived at was 7° 12′, which is 1/50th of Earth's total circumference. So as it was 5,000 stadia from Alexandria to Syene, it would be 5,000 x 50 to circle Earth. The final result of 252,000 stadia (Eratosthenes rounded things up a little to account for small errors) now depends on the length of stadium he used. The standard Greek measure of 185 meters (202 yards) taken from the stadium at Olympia back in Greece produces an error of 16 per cent. However, it is more likely that Eratosthenes used the Egyptian stadion of 157.5 meters (172 yards), which would give the result as 39,690 km (24,662 miles), less than 2 per cent out!

10 Wheels Within Wheels

HIPPARCHUS WAS AN ASTRONOMER WORKING ON THE ISLAND OF RHODES IN THE SECOND CENTURY BC. His star catalog was so accurate that he discovered that the constellations were not quite as fixed as once thought.

The complete works of Hipparchus have been lost over the last 2,000 years, and all we know about his work comes from the reports of others. Most of his working life was spent mapping the heavens from his island observatory. His resulting catalog of 850 stars was of breathtaking accuracy considering he used no optical aids. Many of the measurements were probably made using a cross-staff, a long pole with adjustable crossbeams or transoms. The central staff was lined up with a star, while a transom was moved until it met the line of sight to another. The angles of the triangle produced could be used to calculate the relative positions of the two stars.

Hipparchus may possibly have worked for a few years in Alexandria, the center of learning in the ancient world. He is shown here in that city measuring the angles between stars using a cross-staff.

It is not for nothing that Hipparchus is remembered as a founding figure in trigonometry, producing the first tables of the ratios of a right-angle triangle's angles and lengths. He gave every star coordinates on the celestial sphere, like the longitude and latitude used to pinpoint locations on a globe. Hipparchus also ascribed stars with a magnitude, or measure of brightness.

Heavenly precession

Hipparchus found that his star map did not tally with the observations from his long-dead colleagues, stretching back to Babylon. Some key stars had moved slightly over the centuries. Of course, stars are on the move all the time as the celestial sphere rotates around Earth (as Hipparchus would have understood it). However, Hipparchus' measurements related to the equinoxes, two seemingly fixed moments each year. Hipparchus also found that the time between equinoxes varied. His answer for this was that the equinoxes moved, or "precessed" through the zodiac at 1° per century. (We now know that this is due to Earth's axis wobbling due to the gravitational pull of our neighbors, and it happens slightly faster than Hipparchus said, completing a cycle every 26,000 years.)

Eccentric Sun and Moon

Hipparchus also attempted to measure the distance to the Moon and Sun. He found that the Sun was so far away that its distance was incalculable, but he calculated the distance to the moon as 59 times the radius of Earth (the real distance is 60 radii). He also attempted to reconcile the erratic motion of the Sun and Moon that he observed with the smooth circular path around Earth that the Aristotelian view of the Universe dictated. Hipparchus found that the only way to explain why these bodies appeared to slow and speed up was to introduce epicycles and eccentrics, opting for the latter in the case of the Sun, but using a combination of both to explain the Moon's movements. These solutions are often described as "wheels within wheels" for their unnecessary complexity.

ECCENTRIC VERSUS EPICYCLE

An eccentric is a circular path (P) where the center (C) does not correspond exactly with the object (E, Earth in this case) that is being circled. An epicycle (Q) is a small circle that orbits a point (A) on the circumference of a larger circle, known as a deferent. This point is not fixed, but itself moves around its center, which is the fixed position of Earth (E).

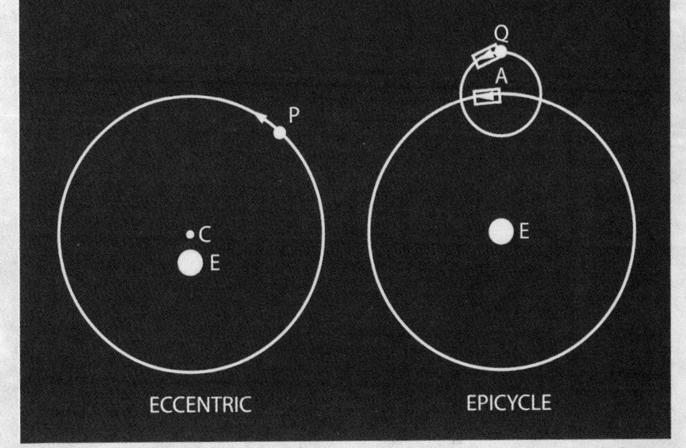

11 Antikythera Mechanism

BROUGHT UP IN 1902 FROM THE SEABED OFF ANTIKYTHERA, A SMALL ISLAND NEAR CRETE THIS INTRICATE SET OF CORRODED dials looked like antique clockwork. Antique indeed—they are from an astronomical computer that is 2,000 years old!

The mechanism was made of brass gears encased in wood. An unknown number of gears are lost, as is the handle that cranked the device.

We know the device was truly ancient because it was found in a wreck off the coast of the isle of Antikythera. The ship was probably filled with the loot of a Roman general returning to Italy around 70 BC. It sunk in 60 meters (198 feet) of water and several divers died in its excavation, which predated SCUBA technology.

The mechanism is built with such precision that experts think that there must have been several earlier versions. They also surmise the gadget came from Rhodes, the home of Hipparchus. The theory goes that the gears were part of a larger mechanical computer that used the orbital systems established by Hipparchus and that could be turned to track the positions of the Sun and Moon with relation to the main stars at any point in time.

12 The Julian Calendar

THE ANCIENT EGYPTIANS HAD BEEN COUNTING YEARS IN SETS OF 365 DAYS FOR CENTURIES, BUT THEY FOUND THAT THEIR CALENDAR gradually went out of sync with the stars. It fell to the great Roman dictator Julius Caesar to solve this astronomical problem.

The 12 months we use today date from the Roman calendar. The Julian reforms insured that each month matched a particular season—November was a time to slaughter livestock.

It appears that the Universe is not in perfect harmony. The time between each sunrise is about four minutes longer than the period between a star rise; the celestial sphere is spinning slightly faster (or so it appears). In addition, Hipparchus and others knew that a year—the time it took for the Sun to move around Earth—was 365 days and 6 hours. That meant that events heralded by certain stars gradually shifted around in the calendar. For example, the dog days of summer was a Roman reference to the rise of Sirius the Dog Star in July. However, by the first century BC the Roman calendar was looking a bit of a mess and extra months had to be inserted rather haphazardly so the date caught up with the actual star time.

When Julius Caesar became leader of Rome in 46 BC, he was determined to sort out the calendar. With advice from astronomer Sosigenes of Alexandria, Caesar introduced a leap year of 366 days every four years (added to Februarius, the shortest month). By this time the old Roman calendar was running three months ahead, so it was decreed that the year now called 46 BC would have two extra months, making it 445 days long, bringing everything back into line.

13 Ptolemy's *Almagest*

CLAUDIUS PTOLEMY WAS THE LAST ASTRONOMER OF NOTE IN THE CLASSICAL ERA. He is remembered for his compilation of contemporary science—riddled with misconceptions but definitely food for thought for those who followed.

Better known simply as Ptolemy, this astronomer was based in Alexandria, a great port at the mouth of the Nile established by the Greek conqueror Alexander the Great (it was not the only city he named for himself). The name Ptolemy invokes a noble line that leads back to Alexander's generals, one of whom took control of Egypt after the diminutive empire-builder had died. The Ptolemies ruled as pharaohs for 250 years until Queen Cleopatra VII became embroiled with a couple of Romans... So, by the time Claudius P. was working in 140 AD, he was a thoroughly Roman citizen with only tangential links to the ex-ruling family.

Almagestũ CL. Ptolemei Pheludiensis Alexandrini Astronomoꝝ principis: Opus ingens ac nobile omnes Celoꝝ motus continens. Felicibus Astris eat in lucē: Ductu Petri Liechtenstein Coloniēsis Germani. Anno Virginei Partus. 1515. Die. 10. Ja. Venetijs ex officina eiusdem litteraria.

* * *

Cum priuilegio.

The title page of a Latin edition of the Almagest *printed in Venice in 1515, in those days still very much a valid astronomical compendium. Less than 30 years later, the work of Nicolaus Copernicus would transform this grand book into a curiosity of history.*

Not all his own work

Ptolemy was a working astronomer and made some helpful advances in trigonometry, but his 13-volume work is largely based on Hipparchus' star catalog and mathematical interpretation of the motion of the Sun and Moon. Ptolemy presented his book as a way of predicting astronomical events, not a representation of the Universe, yet it had the effect of entrenching the Aristotelian system at the dawn of Christendom, making it the official *received* knowledge of the inner workings of creation.

Greek-speaking Ptolemy originally titled the book *Mathematika Syntaxis* (the Mathematical Treatise). It was later repackaged as the *Mega Syntaxis* (the Great Treatise). As the center of astronomy and other sciences moved to the Middle East in the sixth century during the so-called Dark Ages of Europe, the book became known simply as the "Greatest", or *al-majisti* in Arabic. When reintroduced to Europe centuries later, Ptolemy's book was known simply as the *Almagest.*

A clay frieze of Ptolemy shows the astronomer measuring the altitude of a star with a simple quadrant.

The Ptolemaic contribution

Ptolemy did make some extensions to the Hipparchus star catalog and did a lot of work making the observed motion of the Sun and Moon tally with their reputed positions in orbit around Earth. He extended this system to the planets, which added a new level of complexity—the outer planets not only sped up and slowed down, but also appeared to change direction completely when viewed from Earth. Although relatively easy to explain with a heliocentric model, Ptolemy refined (with great mathematical skill) the epicycles and eccentrics of Hipparchus, and introduced equants, a third point within an eccentric about which the center of an epicycle moved.

FINDING EARTH'S PLACE

14 The Astrolabe

DURING THE FIRST MILLENNIUM THE FOCUS OF ASTRONOMERS WAS SEEING THE HEAVENS BETTER AND MEASURING ITS FEATURES MORE ACCURATELY. The instrument of choice was the astrolabe, credited as Hipparchus's invention but developed to its full potential by Islamic astronomers.

As the Roman Empire receded from the fifth century onwards, the center of astronomy moved to the Islamic empire. Baghdad's House of Wisdom became an Arab response to the long-lost Library of Alexandria, and a template for the future universities of Europe.

The main focus of Islamic astronomy was to find ways of measuring the *quiblah*, the direction to the holy city of Mecca, and to explore methods of telling the time, so the faithful across the enormous caliphate could pray at the appointed times in the correct way. To do this astronomers used an astrolabe (literally "star taker"), which was an adjustable two-dimensional model of the sky. An intricate *rete* showed the ecliptic and several bright, easily seen stars. This overlaid a plate showing the coordinates of the celestial sphere. Both moved within an outer frame divided into the hours and minutes of the day. To tell the time, the user measures the altitude of a bright star and adjusts the device to match. Surveyors would have used astrolabes to position the *mihrab*, the niche in a mosque wall that indicates the direction of prayer.

This brass astrolabe was made in Cairo in the 13th century. The inner circle is the ecliptic, while the brightest stars are shown as points on ornate molded constellations. Adjusting these features to match the grid on the back plate gives a representation of the sky.

The Golden Age of Islam also produced its fair share of discovery. The 10th-century scientist Ibn al-Haythem (or Alhazen) answered one of the fundamental questions: what do we see? He showed that light travels from objects (he was especially referring to distant stars) into our eyes, not out of the eyes first, before bouncing back from the surroundings as Ptolemy proposed. The Earth stayed at the center of the Islamic world, with al-Sufi and others updating the work of Ptolemy, adding the first record of the Andromeda galaxy (marked as a "little cloud") and more stars. As a result of their endeavors many of the terms used in modern astronomy have Arabic roots: zenith, azimuth, nadir, and almanac. Also stars like Betelgeuse, Rigel, and Altair have Arabic names.

15 The Crab Nebula Appears

IN 1054 AD A NEW STAR APPEARED IN THE SKY, A BRIGHT LIGHT IN THE CONSTELLATION OF TAURUS. Within two years it had faded from view. Chinese astronomers made the best records of this unusual event, which is now known as a supernova. This particular one left a faint cloud of gas behind which was named the Crab Nebula.

The Crab Nebula was named in 1845 by William Parsons, who saw several appendages emerging from the central region, bringing to mind a crab. This modern image shows a lot more detail of the gas cloud which is still expanding at 1,500 kilometers (932 miles) per second nearly 1,000 years after it formed.

Chinese astronomers had made recordings of "guest stars" before, new arrivals in the night sky. Most of these were probably either comets, or what were later termed a nova by the 16th-century Danish astronomer Tycho Brahe. *Nova* means new in Latin, and refers to any star that suddenly grows brighter for a short while. We now know that the guest star of 1054 was actually a supernova, an especially bright body seen after a giant star explodes. This phenomenon was not explained until the 1930s, and SN 1054 (as the Crab Nebula is now classified) is one of the first to be widely recorded.

Seen around the world

Several Chinese astronomers saw the new star, as did others in Japan and Korea. It has also been suggested that a rock painting in Chaco Canyon, New Mexico, was a Native American record of this new light in the sky. By 1056 the light was gone, but in 1731 English astronomer John Bevis saw a faint smudge in its place. He added it to a growing set of hazy astronomical objects known as nebulae (little clouds), that had been discovered little more than a century before.

16 Copernicus Changes the World

THE PHRASE "PARADIGM SHIFT" IS FREQUENTLY USED, PERHAPS A LITTLE TOO FREQUENTLY NOWADAYS. Strictly speaking it means a change in a set of scientific assumptions, and the person who made the biggest paradigm shift in history was Nicolaus Copernicus.

Imagine going to sleep thinking that the Sun and everything else in the Universe was revolving around you on Earth, only to wake up to be told that our planet was just one of several that moved around the Sun.

Although he was not the first to think it, Copernicus was the first to openly propose such a heliocentric theory, and to back it up with astronomical calculations. He had become interested in astronomy while training to be a doctor—it was believed that health was influenced by the stars back then. He went on to teach astronomy in Rome for three years, presenting the complexities of epicycles and eccentrics that described the motion of the bodies around Earth. It is unclear when he began to consider the far simpler heliocentric alternative, but once he returned to Poland, working as a clergyman no less, he began to circulate his somewhat heretical theory. However, Copernicus did not risk trouble with the boss. He entrusted a manuscript to a loyal student who published it when his master was on his deathbed. The book was immediately banned by the Pope, and stayed banned for 300 years.

This simple illustration from Nicolaus Copernicus's book De Revolutionibus Orbium Coelestium *(On the Revolutions of the Heavenly Spheres) showed the six planets (including Earth) in circular orbit around the Sun.*

Nicolaus Copernicus is a national hero in Poland. This statue of him holding a model of the solar system with the Sun at the center stands outside the Polish Academy of Sciences in Warsaw.

17 Tycho Brahe's Observatory

IT MUST BE REMEMBERED THAT UNTIL THE 1600S EVERY STAR EVER CATALOGED WAS OBSERVED WITH THE NAKED EYE (there are an estimated 2,000 stars visible in this way). Tycho Brahe was the last great astronomer of the pre-telescope era, working in the world's first purpose-built observatories.

Tycho Brahe was the Bond villain of astronomers. Not only was he reportedly an unpleasant person, incredibly wealthy and known only by his first name, he also built elaborate lairs of astronomy on an island between Denmark and Sweden. The first was Uraniborg (Castle of the Heavens), which had tall towers for his equipment, a huge basement for conducting his other experiments, and a large garden. However, the cruel Baltic winds made the towers shake, upsetting delicate observations, and Tycho's ambitious apparatus soon outgrew the cramped building. So Tycho built himself Stjerneborg, the Castle of Stars. This was underground with just the instruments above the surface, protected by windbreaks.

This painting by Dutch cartographer Johan Blaeu, of Tycho Brahe's Uraniborg observatory on the island of Hven (now Ven), shows the great man and his lesser assistants within the main observatory. The basement laboratory, middle reception area, and upper observation floor are also clearly set out.

Celestial changes

Tycho rejected the Copernican model because he could not detect any parallax, or apparent shifts in the stars that you would expect to see if Earth was moving. (They are too far away.) Instead Tycho suggested the Sun orbited Earth but the other planets orbited the Sun. Despite these inaccuracies, Tycho's star map was second to none. One of his earliest recordings was his greatest. In 1572, he saw a nova so bright it rivaled Venus. He found no evidence that this new star (SN 1572) was closer than any other, which meant that the celestial sphere was not perfect and perpetual after all, but open to change like everything else.

18 A New Calendar

DESPITE THE CORRECTIONS ADDED BY JULIUS CAESAR THE WESTERN YEAR WAS STILL NOT QUITE 365 DAYS AND 6 HOURS, but 11 minutes shorter. It took centuries for the error to manifest itself, but when it disrupted the date of Easter, the Pope had to act.

Easter is a movable feast calculated according to the lunar calendar. Early Christian leaders took the theme of resurrection and decreed that Easter would be the first Sunday after the full moon that followed the vernal (spring) equinox—a time when nature is stirring back into life after the winter dormancy.

The Christian calendar is filled with significant events prior to Easter, not least Lent, the period of abstinence, and so it was important to know its date in advance. In calculating the dates clergy had to be careful in identifying the Lent Moon (prior to the equinox). They sometimes resorted to a bit of guesswork and if the moon came too early, they termed it "betrayer," or *belewe* in Old English. An extra moon that appears in a season is now called a blue moon, perhaps from this term.

The method of dating Easter was set at the First Council of Nicaea, the first ever convention of the world's Christian leaders held in 325 AD under the auspices of Emperor Constantine, seen center stage and suspiciously Christ-like in this 15th century depiction. It fell to Pope Gregory 1,250 years later to tweak the process, making it fit for the future.

BLUE MOONS, HARVEST, AND HUNTER'S TOO

The months of the year are approximations of the lunar cycle—an approximation because the phases of the Moon take about 29 days, meaning there are 13 full moons every year. A blue moon is the second full moon in a month—quite a rare event that happens "once in a blue moon." A harvest moon appears around the autumnal equinox, rising soon after the sun has set, providing bright moonshine that illuminates the fields so farmers can continue the harvest late into the night (above). The next full moon, the hunter's moon, also fills the evening sky, often looming large close to the horizon. It is so named because its light was ideal for hunting birds migrating through in the fall, and signals a time of plenty and feasting.

Julian time lag

Those 11 minutes missing from the Julian year meant that the calendar became one day behind the true date every 128 years. By the end of the European medieval period in the sixteenth century, the Julian calendar had slipped by more than a week, which meant Easter was shifting gradually earlier, in sync with the Sun but out of kilter with the calendar. Conversely the fixed feast days, such as Christmas, were moving forward against solar time, very slowly moving out of their traditional seasons. If nothing was done, Easter and Christmas would eventually be on the same day (albeit many millennia away).

In 1578, Pope Gregory XIII decided to act. He took expert advice, mainly from Christopher Clavius, a German mathematician and staunch geocentrist, and announced a small tweak to the Julian system of leap years, so that century years (1600, 1700, etc) would not be leap years unless they were also divisible by 400. This had the effect of reducing the average year to 365 days, 5 hours, 49 minutes and 12 seconds. Gregory wanted to pull the calendar back to an earlier position so Easter occurred in spring. He announced that October 4, 1582, would be followed immediately by October 15. (An 11th day was later removed as well.)

Universal calendar, eventually

Gregory's simple adjustments mean that the year will not fall one day behind until the year 3719. However, it took 350 years for the calendar to be universally adopted. It was implemented in Catholic countries in 1582, but Protestant states were slower to join in. Sweden removed 11 days gradually over 40 years, meaning it was using a unique dating system for all that time. The British Empire (including North America) made the changes in 1752. Meanwhile Turkey carried on using the Julian calendar until 1929.

19 A Magnetic Planet

WE HAVE WILLIAM GILBERT TO THANK FOR THE WORD *ELECTRICITY*, WHICH HE COINED FROM THE GREEK WORD FOR "AMBER." However it is the related phenomenon of magnetism, and its relation to our planet, for which Gilbert is most remembered.

It was the Father of Science himself, Thales of Miletus from sixth-century BC Greece, who first addressed the subject of electricity and magnetism. The former is related to a property of amber, fossilized tree resin. Rubbing a piece of amber gives it a static electric charge which attracts lightweight objects, such as feathers or dust. The magnetic phenomenon is related to *magnítis líthos*—the stones of Magnesia (in central Greece). They are better known now as lodestones, naturally magnetic lumps of iron oxide.

William Gilbert's contribution came 2,200 years later, with *De Magnete* (On the Magnet), published in 1600. In it he reveals that our whole planet is a magnet. Just as the opposite poles of two magnets attract each other, so the points of a compass needle are pulled towards Earth's poles, showing the directions north and south. Gilbert proved this to be the case using a terrella, a model world carved from a lodestone. A compass placed on the surface of the terrella behaved just as it would when used on Earth itself for navigation. The source of the planet's magnetism is theorized to be a spinning core of solid iron, although no one has been to the center of the Earth to check.

An illustration from De Magnete *shows how William Gilbert made magnets by orienting hot iron north-south and then tapping it with a hammer.*

A primitive field diagram from De Magnete *shows the magnetic field at various points on Earth (the north- south axis runs diagonally right to left). It also shows the Earth's magnetism spreading out into space in an Orbis Virtutis, or sphere of power. Direct evidence of this was not collected until the launch of the first artificial satellites in the 1950s.*

20 Lippershey's Telescope

The precise origin of the telescope is unclear. Legend has it that the telescope was really invented by Hans Lippershey's children.

AN ASTRONOMER IS LIMITED BY HIS ABILITY TO DETECT EVIDENCE FROM SPACE. A simple innovation by a Dutch lensmaker offered a chance to see further and in more detail than ever before.

Hans Lippershey lived in Middleburg, a provincial capital in the Netherlands. Lippershey was a lens grinder, shaping discs of glass by hand for use in reading glasses. He had arrived in Middleburg from his native Germany in 1594, a few years after Zacharias Janssen, another lens grinder from the city, made the first microscope by fixing two lenses inside a tube to magnify objects viewed close up. In 1608, Lippershey built a scaled up device that did something similar to distant objects. Legend has it that Lippershey's children discovered that holding two lenses a certain distance apart magnified the weather vane on the nearby church. It is a distinct possibility that he took the idea from Janssen, who worked nearby and produced a spyglass around the same time. Whatever the truth, Lippershey's attempts to profit from the "Dutch perspective glass" failed because it was widely copied.

21 Kepler's Laws of Planetary Motion

JOHANNES KEPLER WAS BY NO MEANS FREE OF SUPERSTITIONS—HE WAS A PROFESSIONAL ASTROLOGER AS WELL AS A GIFTED MATHEMATICIAN—but he was the right man at the right place with the right data, and found the first truly scientific universal law of astronomy.

Religious persecution in southern Germany and Austria forced Kepler, a Lutheran, to move to Prague, where he became Tycho Brahe's assistant. The Dane had run out of friends in his home country and had become the court astronomer for Emperor Rudolph, the Holy Roman Emperor. Rudolph was an inept politician, and his reign led

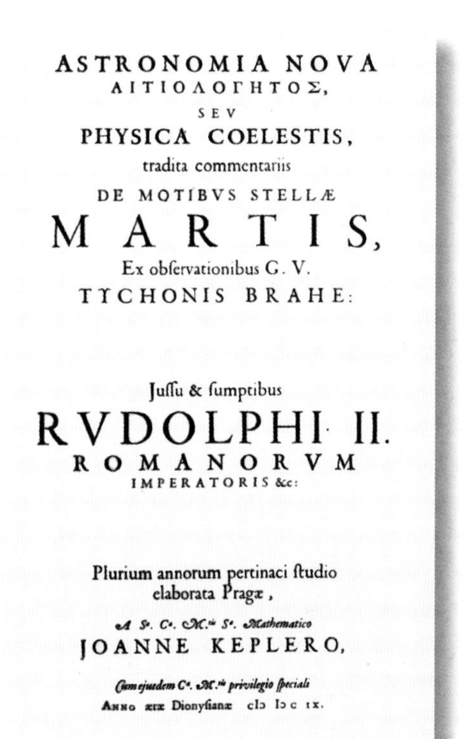

ASTRONOMIA NOVA
ΑΙΤΙΟΛΟΓΗΤΟΣ,
SEV
PHYSICA COELESTIS,
tradita commentariis
DE MOTIBVS STELLÆ
MARTIS,
Ex obſervationibus G. V.
TYCHONIS BRAHE:

Juſſu & ſumptibus
RVDOLPHI II.
ROMANORVM
IMPERATORIS &c:

Plurium annorum pertinaci ſtudio
elaborata Pragæ,

A S. C. M. S. Mathematico
JOANNE KEPLERO,

Cum ejusdem C. M. privilegio ſpeciali
Anno æræ Dionyſianæ cIↄ Iↄ c IX.

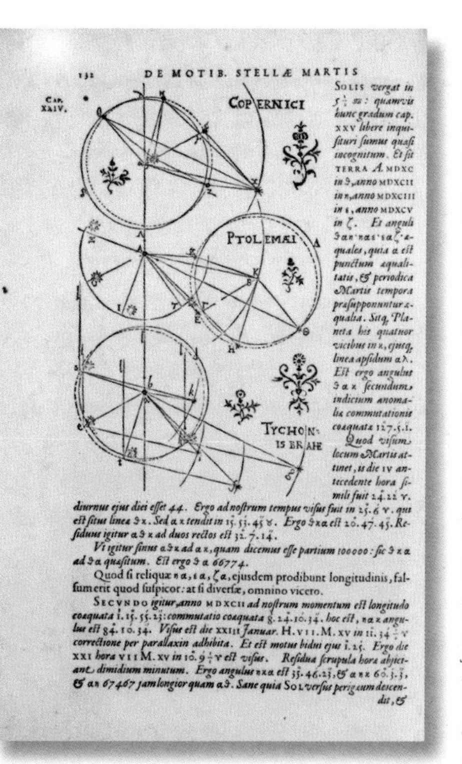

With these sketches from Astronomia Nova *Kepler makes a comparison between descriptions of the Universe made by Copernicus, Ptolemy, and Tycho.*

to the German-speaking world being shattered by the Thirty Years' War. However, he was an avid patron of the arts and sciences whose influence was one of the seeds of the coming Scientific Revolution.

After Tycho's death in 1601, Kepler was bequeathed his records of planetary motion, which the Dane had jealously guarded in life. Unlike Tycho, Kepler was a Copernican, yet like centuries of astronomers before him, he assumed that the planets turned in perfect circles. The best data was for Mars, and Kepler spent six years studying them. His conclusions were set out in 1609 in the boldly titled *Astronomia Nova* (The New Astronomy). Planets did not *circle* the Sun at all; the only thing that made sense was that they moved in ellipses.

Kepler subscribed to the ancient Greek concept that the Universe was mathematically harmonized. Before meeting Tycho he attempted to find a way of organizing the orbits of the six known planets as spheres inscribed inside and out of the five Platonic solids. He published his system in Mysterium Cosmographicum *(The Cosmic Mystery) in 1596.*

Change of focus

The geometry of the ellipse had been well understood since the classical era of ancient Greece, belonging to the family of curves known as the conic sections, formed by a plane cutting through a cone. An ellipse has two focal ponts or foci, and the sum of the distances from any point on the curve to the foci is constant. A circle is just a special ellipse with a single focus.

Kepler summed up planetary motion in three mathematically rigorous laws: the first simply, states that planets orbit in ellipses, the second describes how planets move faster in areas of the orbit that are closer to the Sun, and slower when further away. In math terms, a line between the planet and the Sun sweeps out the same area per unit of time wherever you measure it or for how long. Finally, the third law is that the square of the orbital period (the planet's year) is proportional to the cube of the average distance to the Sun.

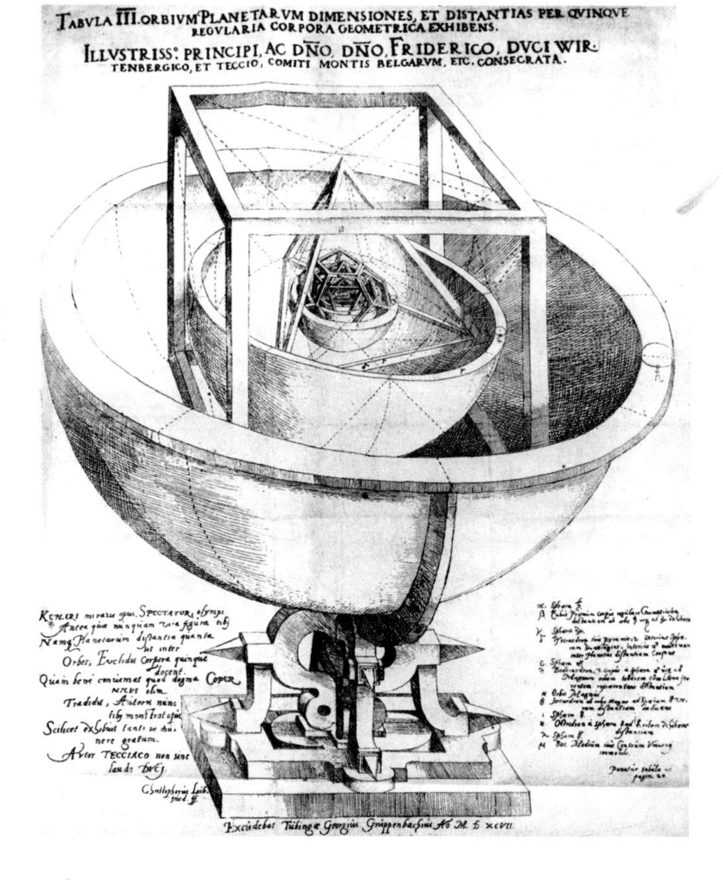

22 The Starry Messenger

GALILEO GALILEI WAS SUCH AN ILLUSTRIOUS SCIENTIST THAT HISTORIANS HAVE DONE AWAY WITH HIS LAST NAME. In 1610, the Italian turned a telescope—homemade but still the world's best at the time—to the heavens. What he saw turned astronomy on its head—and landed him in a lot of trouble.

Galileo is best remembered for his astronomical observations and vocal—although inconsistent—support for heliocentrism, the school of thought that moved Earth into orbit around the Sun and into direct conflict with the teachings of the all-powerful Catholic Church. However, Galileo had already been chipping away at the Aristotelian Universe (to which the Church subscribed as a revealed truth) in his previous work. In 1589, legend has it that he dropped two iron balls—one much heavier than the other—from the Leaning Tower of Pisa. He found that they fell "evenly," reaching the ground at the same time. This was contrary to Aristotle's theory of gravitation, which stated that the heavy one should fall faster than the smaller.

Light and money

However, Galileo found that a life of science was not treating him in the way in which he would like to be accustomed. Upon hearing about Hans Lippershey's invention and

Galileo shows off his telescope to dignitaries of Venice. His most powerful device had a magnification of x 30, ten times the power of Lippershey's.

THE REFRACTING TELESCOPE

Galileo made his astronomical discoveries using a refracting telescope. This uses two lenses to focus and magnify the light of the stars. The lensing effects of curved glass had been understood for centuries. Light refracts, or changes direction, as it crosses into glass and out again. The curve of a lens changes the angle of refracted light running through it, so all beams are focused on one point on the other side. This is what the objective lens —the large one at the front—of a telescope does: gather light and focus it into a small but intense image. The role of the smaller lens, the eyepiece, is to magnify the objective image.

The second lens diverges the light coming from the objective image, so it appears to the viewer as a larger "virtual" image, located slightly further away, and visible in greater detail than the original object.

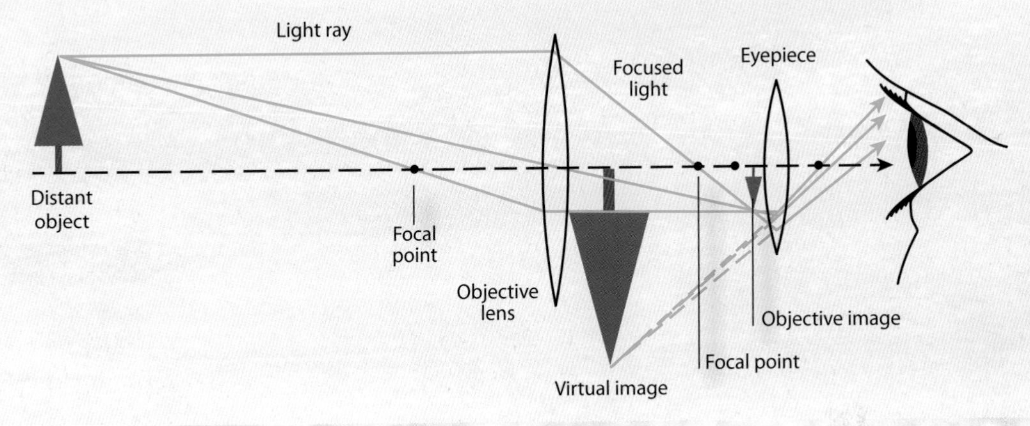

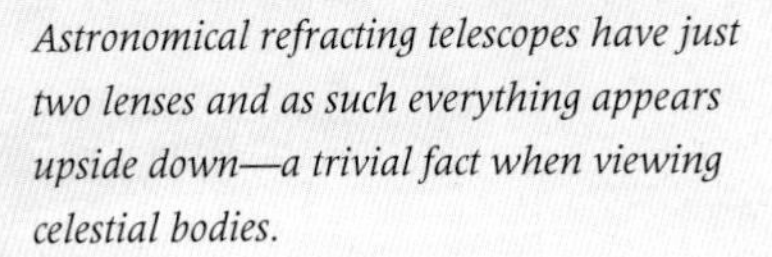

Astronomical refracting telescopes have just two lenses and as such everything appears upside down—a trivial fact when viewing celestial bodies.

realizing that telescopes would soon be on sale in Venice, Galileo formulated a get-rich-quick scheme. Such a device would be enormously valuable for the great maritime city; signals from approaching ships relaying cargo details were enough to adjust prices in Venetian markets. Anyone equipped with a telescope would have the upper hand. The Venetians instructed Galileo to retroengineer one rather than pay the Dutch inventor. Galileo's version was even more powerful and at a stroke his academic's salary was doubled in reward.

Into the night

Galileo was not the first person to use a telescope to look at heavenly bodies, but his 1610 book—*Sidereus Nuncius* (The Starry Messenger)—is the first scientific report on the subject. The great astronomer saw that Venus, the brightest object in the sky, went through phases just like the Moon. That meant the planet's illuminated side was often only partially visible from Earth. How could that be if it orbited Earth as Church dogma dictated? The evidence pointed to an Earthly orbit around the Sun. Turning his attention to Jupiter, Galileo saw the planet surrounded by three stars. Each night these "stars" moved, and a few days later a fourth appeared. Galileo realized he was looking at four moons that were orbiting Jupiter. But in Aristotle's Universe everything moved around Earth.

As news spread of these discoveries, Galileo found his case before the Inquisition (the guardians of Church dogma). Theologians declared the idea that the Sun was a stationary object to be plainly absurd, and Galileo was ordered to recant his theory. He resisted doing so for several years, but eventually faced trial for heresy and was forced to back down to avoid jail. In 1992, the Vatican apologized for the mistreatment of Galileo.

OBSERVAT. SIDEREÆ. 19

Hæc eadem macula ante ſecundam quadraturam nigrioribus quibuſdam terminis circumvallata conſpicitur, qui tanquam altiſſima montium juga ex parte Soli averſa obſcuriores apparent, quà vero Solem reſpiciunt, lucidiores extant. cujus oppoſitum in cavitatibus accidit, quarum pars Soli averſa ſplendens apparet. obſcura verò ac umbroſa, quæ ex parte Solis ſita eſt. Imminuta deinde luminoſa ſuperficie, cum primum tota ferme dicta macula tenebris eſt obducta, clariora montium dorſa eminenter tenebras ſcandunt. Hanc duplicem apparentiam ſequentes figuræ commonſtrant.

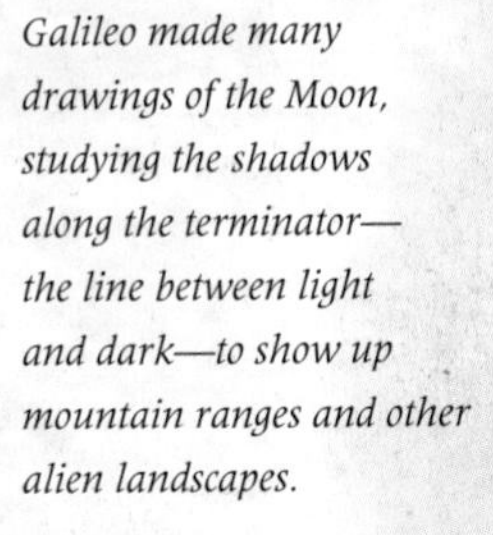

Galileo made many drawings of the Moon, studying the shadows along the terminator—the line between light and dark—to show up mountain ranges and other alien landscapes.

23 A Transit of Venus

KEPLER DID NOT LIVE TO SEE HIS MATHEMATICAL DESCRIPTION OF PLANETARY MOTION PUT TO USE. But soon after Kepler's death an amateur astronomer used the German's laws to show that the orbit of Venus would pass right in front of the Sun in 1639.

In 1859 a stained glass window was installed at St. Michael's Church at Hoole, Lancashire, depicting Jeremiah Horrocks, the one-time curate, observing Venus transiting the Sun. He is shown using a sheet instead of a card to display the bright image.

Jeremiah Horrocks was the curate of a church in rural Lancashire but he had received a thorough grounding in the work of Copernicus, Galileo, and Kepler while at Cambridge University. The most widely circulated tables of the orbit of Venus had been made by Dutchman Philippe van Lansberge, and Kepler himself had used them to predict that Venus would be passing very close to the Sun in 1639. Horrocks made his own observations of Venus and was sure that Lansberge's were inaccurate—Venus would actually transit, pass in front of the Sun, on November 24 that year. Horrocks focused the Sun through a telescope onto a card. Sure enough at 3:15 he saw a tiny dot on the surface—the shadow of the planet Venus. Kepler's math was a perfect astronomical tool.

24 Huygens Sees Saturn's Rings

THE VIEW THROUGH EARLY TELESCOPES WAS FAR FROM CLEAR. The hand-crafted lenses frequently distorted the image of the heavens, with the light of different colors focused in different ways. Only once this chromatic aberration was reduced did the details come into view.

Galileo's observations are remembered for changing the big picture, but they are very short on detail. His drawings of lunar features are hard to identify with the real thing—his telescope no doubt gave a highly distorted image. The main point was that Galileo could see for the first time that the Moon was not a smooth orb, but another world with a surface as rugged as Earth's. He also found that Jupiter had four moons of its own, and that Saturn was actually three bodies, one large, two small, that stayed close together... The last one was obviously an artifact of his crude instrument, made all the more confusing when the two outer bulges disappeared from view soon after,

SYSTEMA SATVR... 47

ea quam dixi annuli inclinatione, omnes mirabiles Saturni facies ſicut mox demonſtrabitur, eo referri poſſe inveni. Et hæc ea ipſa hypotheſis eſt quam anno 1656 die 25 Martij permixtis literis una cum obſervatione Saturniæ Lunæ edidimus.

Erant enim Literæ aaaaaaacccccdeeeeegh iiiiiiillllmmnnnnnnnnnoooopp q rr stttt t uuuuu; quæ ſuis locis repoſitæ hoc ſignificant, *Annulo cingitur, tenui, plano, nuſquam cohærente, ad eclipticam inclinato.* Latitudinem vero ſpatij inter annulum globumque Saturni interjecti, æquare ipſius annuli latitudinem vel excedere etiam, figura Saturni ab aliis obſervata, certiuſque deinde quæ mihi ipſi conſpecta fuit, edocuit: maximamque item annuli diametrum eam circiter rationem habere ad diametrum Saturni quæ eſt 9 ad 4. Ut vera proinde forma ſit ejuſmodi qualem appoſito ſchemate adumbravimus.

Cæterum obiter hic iis reſpondendum cenſeo, quibus novum nimis ac fortaſſe abſonum videbitur, quod non tantum alicui cæleſtium corporum figuram ejuſmodi tribuam, cui ſimilis in nullo hactenus eorum deprehenſa eſt, cum contra pro certo creditum fuerit, ac veluti naturali ratione conſtitutum, ſolam iis ſphæricam convenire, ſed & quod annulum

Occurritur iis quæ de annulo objici poſſent.

The rings made their first appearance in Christiaan Huygen's Systema Saturnium *in 1659.*

only reappearing in 1616. It took until 1655, with advances in telescope technology, to show that these bulging "moons" were in fact the famous Saturnine ring system—which is sometimes barely visible when the planet is viewed side on.

The man credited with this discovery was Christiaan Huygens, the great Dutch scientist (who was soon to invent the pendulum clock and internal combustion engine). Huygens reduced the chromatic aberration of his telescope by using a large weak lens; the thinner glass caused less distortion. Huygens realised that the role of this lens was to collect light from the sky, not magnify it, and he constructed long aerial telescopes, dispensing with a tube altogether. The large objective lens was positioned atop a tall pole, while the earthbound astronomer used an eyepiece to inspect the image it formed. Although hard to keep steady, such devices made several discoveries until reflecting telescopes took over.

25 Newton's Reflecting Telescope

A SIMPLE WAY TO REMOVE THE PROBLEM OF CHROMATIC ABERRATION IN TELESCOPE LENSES is to make one that gathers light with mirrors instead.

Newton's first reflecting telescope is lost; his second telescope was gifted to the Royal Society of London.

Soon after the invention of the refracting telescope, Galileo and others had speculated on the possibility of using a curved mirror to form the objective image. They thought that light arriving from the stars could be reflected back into a point which could be magnified with an eyepiece lens. All attempts to make a mirror that could reflect a clear image failed until Isaac Newton in 1668. He ground out the curved surface from an alloy of tin and copper, which polished well. This main mirror was housed at the end of a wooden tube and reflected an image onto a second smaller mirror near to the aperture (the open end of the tube). The image was then redirected into an eyepiece lens on the side of the device. The telescope was what brought Newton (just 26) to the attention of the scientific world—his theory of gravity and calculus were decades away. (The world's most powerful optical telescopes today are still reflectors.)

26 Setting Meridians

THE AGE OF EXPLORATION GAVE WAY TO A STRING OF MARITIME EMPIRES, AND ASTRONOMY SOON BECAME AN IMPORTANT STATE FUNCTION. Astronomers maintained accurate time and provided navigation data for the merchant fleets crossing the oceans. And the techniques used to survey the sky were put to work mapping territorial possessions, new and old.

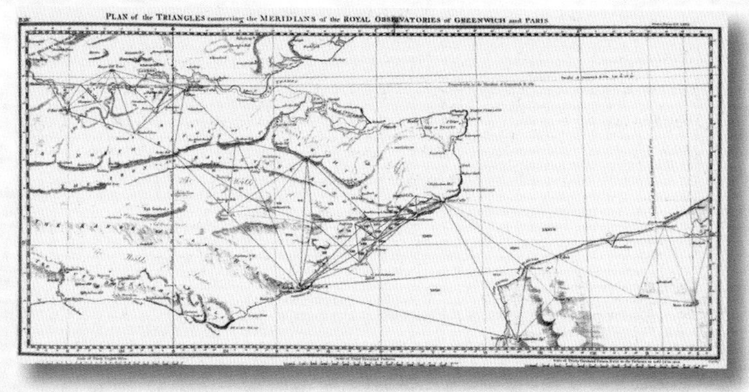

A map of the English Channel (or La Manche in French) shows triangulations connecting the Greenwich and Paris Meridians, allowing maps constructed under the rival systems to be compared.

Every map needs a reference point from where to measure all other points. In terms of a sphere like Earth, such references take the form of great circles that divide the globe in half. The equator is one such circle running halfway between the poles. Every point on Earth has latitude, measured north or south of this line. But there is no similarly natural place to put a meridian—a great circle line running through the poles that can be the reference point for longitude, measurements to the east or west.

In the 1494 Treaty of Tordesillas, Spain and Portugal set a meridian of 370 leagues (1,786km/1,110 miles) west of the Cape Verde islands (a recent Portuguese acquisition). Everything west of the meridian was claimed by the Spanish; the Portuguese got the east. The rich coast of what is now Brazil was in the Eastern Hemisphere, which is why that country has Portuguese heritage among Spanish neighbors.

Paris first, then London

Fifteen centuries earlier Ptolemy had suggested that the most western point of land should be the 0 meridian (since most of the known world lay to the east). In 1634, the French authorities took this as inspiration when setting a meridian at El Hierro, the westernmost Canary Island. In 1667, the meridian was shifted to Paris, running through the center of the brand new Paris Observatory. One of the roles of this

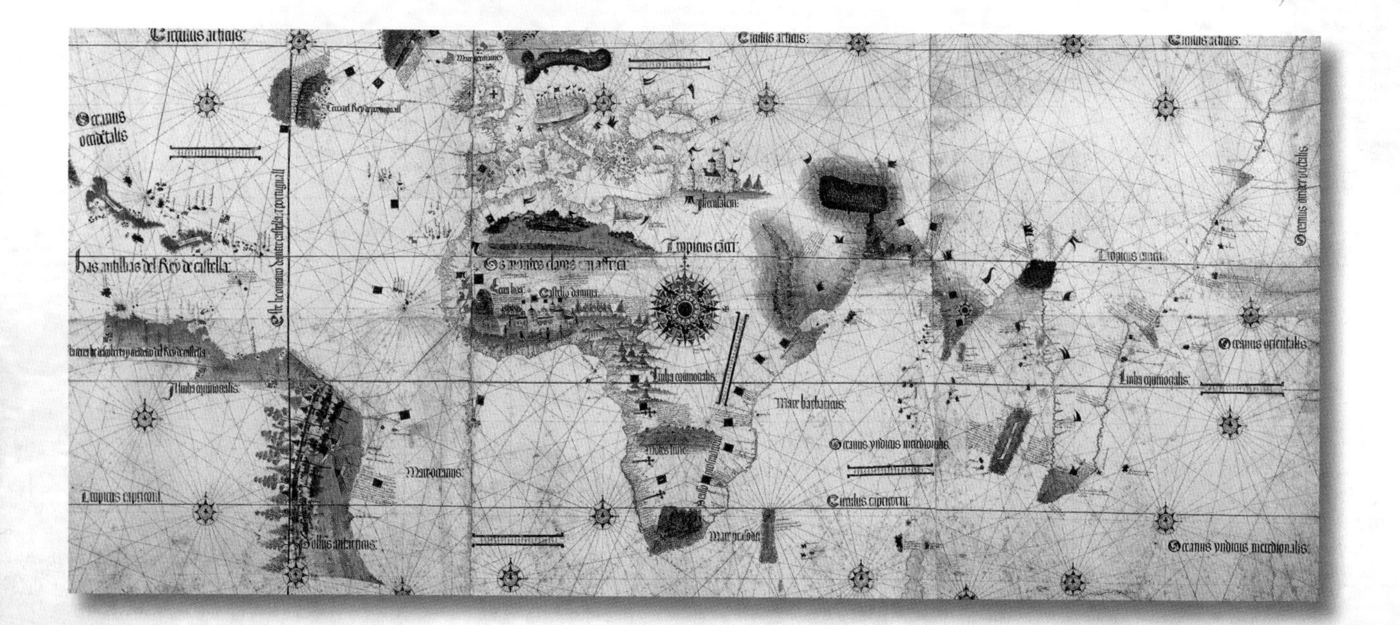

The Octagon Room of the Royal Observatory at Greenwich where John Flamsteed and the other early Astronomers Royal made most of their observations. It has a red time-ball on the roof that is raised at 12.58 and lowered at precisely 1 p.m. each day, originally as a time signal to Londoners and to ships preparing to sail out of the Pool of London in the wide river below.

establishment was to observe the sun passing over the meridian, which indicated the precise moment of noon. (The morning hours are a.m., or ante meridian, while afternoon and evening is p.m., post meridian.) Of course this apparent solar motion was understood by now as due to Earth's rotation. If Earth rotated the full 360 degrees in 24 hours, then it moved one degree in 4 minutes. In 1670 Abbé Jean Picard calculated the size of one degree of longitude as 110.46 km (69 miles).

The British decided they, too, needed an astronomical institute, and the Royal Observatory opened at Greenwich on a hill east of the City of London in 1676. An Astronomer Royal was appointed to run it, the first incumbent being John Flamsteed. He and his successors used a variety of meridians running through the observatory complex. The single Greenwich Meridian used today was not precisely defined until 1851.

27 The Speed of Light

EARLY SCIENTISTS MADE SEVERAL ATTEMPTS TO MEASURE THE SPEED OF LIGHT. However, earthly experiments were confounded by the extraordinary speed. An astronomical solution was needed.

Ole Rømer observing through his telescope with the tools of astronomy around him at the Paris Observatory.

In his dotage Galileo claimed to have tried to measure the speed of light using the time it takes for light from lanterns to travel between observers. He failed to come up with a figure but claimed it was a finite value. Johannes Kepler, on the other hand, thought light filled infinite space instantaneously. In the end it was the Galilean moons of Jupiter that provided the first real answer. In 1676, Ole Rømer, working in the Paris Observatory, compared observations of Io (all four moons had by now been named after Jupiter's various romantic conquests) with its motion predicted by Kepler's laws. Io is hidden when it moves behind Jupiter, but Rømer knew the exact times Io would reappear. However, he found that the moon appeared 10 minutes late and realised that was the time it took for its light to reach Earth. The delays got slightly longer as Earth moved away from Jupiter in its orbit around the Sun. Rømer calculated that light traveled at 220,000 km (136,702 miles) per second, around 25 per cent slower than the actual value.

28 Universal Law of Gravitation

ARISTOTLE HAD SAID THAT LARGER, MASSIVE OBJECTS FELL TO EARTH MORE QUICKLY THAN SMALLER ONES. Galileo had shown that all objects fell "evenly" early in the 17th century around the time when Kepler was wondering what it was that made planets move along their orbits. Was it magnetism? Isaac Newton explained it was all down to a single force called gravity governed by a simple universal law.

Newton suggested that the Moon should be considered as a projectile. A projectile follows a curved path as it is pulled back to earth by gravity. The surface of the Earth is also curved, so if a projectile is traveling fast enough, its curved path will follow the curve of the Earth—and "fall" around the Earth. If the velocity of the projectile is increased the path it takes around the Earth becomes an ellipse.

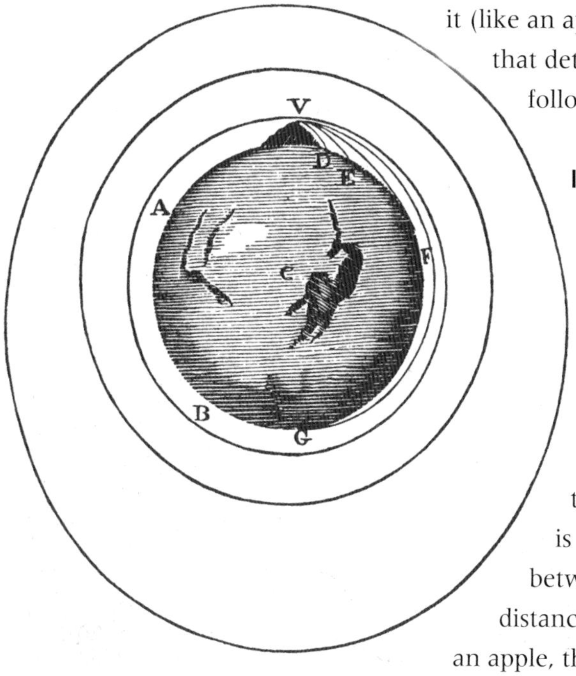

Galileo had shown in experiments that the distance a body falls under gravity is proportional to the square of the time it is falling: a ball falling for two seconds falls four times farther than the same object falling for one. He also showed that a ball's speed was directly proportional to the duration of the fall, and went on to deduce that a projectile fired upwards followed a parabolic path, one of the conic sections, and is thus related to the ellipses of Kepler's planetary motion. What was the connection?

When Cambridge University was closed in the mid 1660s to avoid the Great Plague that was sweeping through England at the time, Newton returned to the family farm, and this was where he formulated his law of gravity, although he kept it quiet for another 20 years. (The falling apple was not part of the story until an octogenarian Newton was telling fire-side stories after a large dinner.) He proposed that everything in the Universe produces a gravitational force that pulls other objects towards it (like an apple falling to the ground) and it is also the force of gravity that determines the paths that the stars, planets, and all other objects follow through space.

Inverse square

Newton was able to calculate the rate of acceleration of the Moon under Earth's gravity and found it to be thousands of times smaller than that of something closer to home (that apple again). How could this be if they were being acted upon by an identical force? Newton's explanation was that the force of gravity was weakened by distance. An object on Earth's surface is 60 times closer to the center of the Earth than the Moon is. The Moon experiences a force of gravity that is $1/3600$ or $1/(60)^2$ that of the apple. Thus the force of gravity between two objects is inversely proportional to the square of the distance that separates those objects, whether they are the Earth and an apple, the Earth and the Moon, or the Sun and a comet. Double the distance and the force is reduced to a fourth, triple the distance and the force reduces to a ninth, and so on. The force is also determined by the masses of the objects—the greater the mass, the greater the force of gravity. Putting it another way, $F = G(Mm / r^2)$. The force of gravity equals a large mass M multiplied by a small mass m divided by the square of the distance between them r^2, all multiplied by the gravitational constant G. The constant, known as "Big G" ("Little g" is the acceleration due to gravity,) determines the attraction between any two masses anywhere in the universe. It was used to calculate the weight of Earth in 1789 (6×10^{24} kg/13.2×10^{24} lbs). Newton's laws work well to predict the forces and motion of two objects. What happens when you add one more is much more complex, and led (300 years later) to the math of chaos.

29 Halley's Comets

[7]

The Astronomical Elements of the Motions in a Parabolick Orb of all the Comets that have been hitherto duly observ'd.

Comet. An	Nodus Ascend. gr. ′ ″	Inclin. Orbitæ. gr. ′ ″	Perihelion. gr. ′ ″	Distan. Perihel. à Sole.	Log. Dist. Perihelii à Sole.	Temp. equat. Perihelii. d. h. ′	Perihelion à Nodo. gr. ′ ″	
1337	♊ 24.21. 0	32.11. 0	♉ 7.59. 0	40666	9.609236	June 2. 6.25	46.22. 0	Retrog.
1472	♑ 11.46.20	5.20. 0	♉ 15.33.30	54273	9.734584	Feb. 28 22.23	123.47.10	Retrog.
1531	♉ 19.25. 0	17.56. 0	♒ 1.39. 0	56700	9.753583	Aug. 24.21.18½	107.46. 0	Retrog.
1532	♊ 20.27. 0	32.36. 0	♋ 21. 7. 0	50910	9.706803	Oct. 19 22.12	30.40. 0	Direct.
1556	♍ 25.42. 0	32. 6.30	♑ 8.50. 0	46390	9.666424	Apr. 21.20. 3	103. 8. 0	Direct.
1577	♈ 25.52. 0	74.32.45	♌ 9.22. 0	18342	9.263447	Oct. 26 18.45	103.30. 0	Retrog.
1580	♈ 18.57.20	64.40. 0	♋ 19. 5.50	59628	9.775450	Nov. 28 15.00	90. 8.30	Direct.
1585	♉ 7.42.30	6. 4. 0	♈ 8.51. 0	109358	0.038850	Sept. 27 19.20	28.51.30	Direct.
1590	♍ 15.30.40	29.40.40	♏ 6.54.30	57661	9.760882	Jan. 29. 3.45	51.23.50	Retrog.
1596	♒ 12.12.30	55.12. 0	♏ 18.16. 0	51293	9.710058	Juli 31.19.55	83.56.30	Retrog.
1607	♉ 20.21. 0	17. 2. 0	♒ 2.16. 0	58680	9.768490	Oct. 16. 3.50	108.05. 0	Retrog.
1618	♊ 16. 1. 0	37.34. 0	♈ 2.14. 0	37975	9.579498	Oct. 29 12.23	73.47. 0	Direct.
1652	♊ 28.10. 0	79.28. 0	♈ 28.18.40	84750	9.928140	Nov. 2.15.40	59.51.20	Direct.
1661	♊ 22.30.30	32.35.50	♋ 25.58.40	44851	9.651772	Jan. 16.23.41	33.28.10	Direct.
1664	♊ 21.14. 0	21.18.30	♌ 10.41.25	102575	0.011044	Nov. 24.11.52	49.27.25	Retrog.
1665	♏ 18.02. 0	76.05. 0	♊ 11.54.30	10649	9.027309	Apr. 14. 5.15½	156. 7.30	Retrog.
1672	♑ 27.30.30	83.22.10	♉ 16.59.30	69739	9.843476	Feb. 20. 8.37	109.29. 0	Direct.
1677	♏ 26.49.10	79.03.15	♌ 17.37. 5	28059	9.448072	Apr. 26.00.37½	99.12. 5	Retrog.
1680	♑ 2. 2. 0	60.56. 0	♐ 22.39.30	00612	7.787106	Dec. 8.00. 6	9.22.30	Direct.
1682	♉ 21.16.30	17.56. 0	♒ 2.52.45	58328	9.765877	Sept. 4.07.39	108.23.45	Retrog.
1683	♍ 23.23. 0	83.11. 0	♊ 25.29.30	56020	9.748343	Juli 3. 2.50	87.53.30	Retrog.
1684	♐ 28.15. 0	65.48.40	♏ 28.52. 0	96015	9.982339	Mai 29.10.16	29.23.00	Direct.
1686	♓ 20.34.40	31.21.40	♊ 17.00.30	32500	9.511883	Sept. 6.14.33	86.25.50	Direct.
1698	♐ 27.44.15	11.46. 0	♑ 0.51.15	69129	9.839660	Oct. 8 16.57	3. 7. 0	Retrog.

This Table needs little Explication, since 'tis plain enough from the Titles, what the Numbers mean. Only it may be observ'd, that the *Perihelium* Distances, are estimated in such Parts, as the Middle Distance of the Earth from the Sun, contains 100000.

A

The table of comet data from Edmond Halley's A Synopsis of the Astronomy of Comets *of 1705 showed that comets orbited the Sun, only in very eccentric orbits.*

THE WORD *COMET* MEANS "LONG-HAIRED STAR," A GREEK VIEW OF THESE RARE and mysterious visitors to the night sky. In 1705, an English astronomer made a name for himself (and a comet) using Newton's laws of gravity.

Comets would have been hard to miss in the prehistoric night—much more obvious than in today's electric-lit sky. Aristotle tackled them in his book on meteorology, suggesting they were atmospheric rather than astronomical. This idea stuck for a long time—beyond the Copernican Revolution. Even when Tycho Brahe had showed that comets were passing by far beyond the distance of the Moon, the great Kepler and Galileo refused to accept that these were bodies controlled by the same laws as the planets.

However, that is exactly what English astronomer Edmond Halley thought. When putting together a list of comets and their orbits, based on observations stretching back to the 1300s, he noticed that three comets seen in 1531, 1607, and 1682 had the same orbit. He applied Newton's laws of motion and gravity to the dates to predict that they were actually the same object that orbited the Sun every 76 years (passing Earth briefly as it approached). It duly returned in 1758 and has been known as Halley's Comet ever since.

UNLUCKY (FOR SOME)

Perhaps because they were not confined to the zodiac, the appearance of a comet is frequently viewed as heralding some kind of a drastic tumult, which is rarely awaited with eagerness. Comet arrivals rarely pass without comment, and the 1066 visit of Halley's Comet is recorded in the Bayeux Tapestry (below.) The words say: "These men wonder at the star," the suggestion being it is a warning to the English that the Normans would soon be invading. Of course, the Vikings also attacked shortly before them at the other end of the country, so it was a really bad year for the English.

30 The Shape of Earth

In his treatise on the shape of Earth, Aristotle made reference to how an object collapsing into its center would form a sphere. By the 18th century, professional astronomers agreed that the centrifugal force created as Earth spun made the planet bulge. But in which direction?

The 1736 French mission to Lapland battled sub-zero temperatures to survey a north-south meridian. Around 60 years later, the quarter meridian from the North Pole to the Equator (through Paris) was used to define the new metric unit of the meter. The length of the arc was set as 10 million meters (11 million yards).

Christiaan Huygens had surmised that Earth was flattened at the poles like an orange—the mathematical term used was oblate spheroid—and the gravitational calculations of Isaac Newton supported this view. René Descartes thought the opposite, and a hands-on survey by Jacques Cassini, the director of the Paris Observatory (having taken over from his better remembered father Giovanni), found that the distances covered by one degree of latitude (moving northward through France) appeared to increase. That suggested that Earth was more like a lemon, taller than it is round (prolate in the jargon).

It was crucial to find out not just for scientific purposes but also to ensure that maps, and the lines of latitude and longitude marked on them, were accurate representations of the surface of the planet. That meant having accurate measurements for Earth's polar and equatorial circumferences. They were not quite the same, and

whichever was the larger, both could be used to calculate a usable average distance for an arc of one degree.

Geodesic missions

What was needed was a new science, geodesy, the study of the shape of Earth. To answer the fundamental question of this new endeavor, French King Louis XV sent out two missions. The first was to measure a meridian arc at the equator. In 1735, a team left for the Spanish territory of Quito (now Ecuador, a name that literally means *equator* in Spanish) taking a total of four years to get back to France with their results. In the meantime another team, including the Swede Anders Celsius (prior to making his name with his centigrade temperature scale), went to Lapland, close to the North Pole, where they measured a similar length of arc as the equatorial mission. Both results clearly showed that Huygens and Newton were correct. We live on an oblate planet.

31 Mapping the Southern Sky

NICOLAS LOUIS DE LACAILLE WAS NOT THE FIRST ASTRONOMER TO MAKE A SURVEY of the sky south of the equator, but he certainly left his mark.

Geodesic data gave a idea of the shape of the northern hemisphere, but what of the southern side? To measure a southern meridian arc was one of the goals of Frenchman Nicolas Louis de Lacaille's mission to South Africa in the early 1750s. His measurements suggested that Earth was egg-shaped (pointed side down), but this was later shown to be an error. However, Lacaille's trip is now remembered for his star survey which added 14 constellations to the sky (more than any person before or since). A thorough Enlightenment man, Lacaille saw the tools of art and science in the heavens, not figures from myths. His constellations include a telescope, clock, easel, chisel, and Antlia, the air pump, the device invented by Denis Papin and utilized by Robert Boyle to study the properties of gases the century before. The results of Boyle's experiments began to reveal the nature of matter on Earth—and out in space.

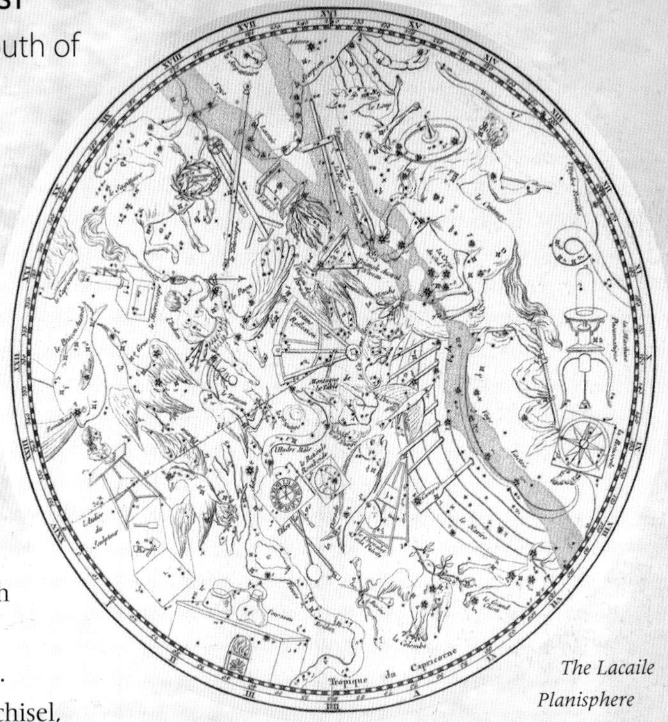

The Lacaile Planisphere extends the ancient constellations of myths and monsters south, showing more modern patterns.

32 Navigating with Astronomy

NAVIGATING BY THE STARS WAS NOT A SCIENTIFIC DISCOVERY. ANCIENT MARINERS HAD EXPERT KNOWLEDGE OF WHICH CONSTELLATIONS DIRECTED THEM TO DIFFERENT DESTINATIONS—MORE OR LESS. However, it took scientific innovation and industrialized production for the heavens to reveal any point on the planet: land or sea, at any time of day or night.

This 17th-century explorer is using a backstaff and compass to measure heavenly bodies. This delicate equipment could only be used accurately on land.

In the ancient world navigation, although not without its risks, was generally limited to hugging coastlines (or perhaps criss-crossing narrow seas such as the Mediterranean Sea). Long voyages were restricted to the most favorable season, and a captain could tell the general direction he needed from the rising and setting of the Sun and by keeping certain constellations in view. It was frequently a bit hit and miss, but a ship was seldom far from land and even if it made landfall off the intended course, it was easy enough to figure out a new route.

Some seamen ventured further out, of course. The Polynesian culture was spreading through long-distance island hopping, relying on charts made from bound sticks, which showed the ocean currents connecting bits of land. The stakes were high, with many misses for every hit.

Scientific advance

In 1418, Prince Henry of Portugal established a school of navigation. To Portugal, out on the western fringe of Europe, the Atlantic Ocean was a gateway to new opportunities that the prince, now remembered as Henry the Navigator, was keen to exploit with improved ship technology and navigation techniques. By then the compass had been in use for more than 200 years (longer in China) but it did not measure distances.

From Aristotle to Eratosthenes, ancient astronomers had often referred to how the altitude—measured as angles above the horizon—of prominent stars changed on long journeys to the north or south. They would have known of this phenomena from the reports of seamen who, for example, saw the North Star (strictly speaking called Polaris nowadays) rise higher as they sailed north from Alexandria to the Aegean cities, and sink lower on the return journey. What was needed was a way of linking these altitudes to a specific latitude. Astronomical records, originally meant for use in astrology, were transformed by Islamic navigators into almanacs—a record of the positions of fixed and wandering stars throughout the year.

The first sextant was made in 1757 by the master British instrument maker John Bird. It was named because its arc covers just a sixth of a circle (60°). The device was a refinement to a smaller navigational tool called the octant (an eighth of a circle) invented in the 1740s. (Isaac Newton is reputed to have invented one in 1699 but never told anyone.)

Getting the right angle

Astronomical tools, such as the cross-staff and astrolabe, were repurposed for use in navigation. Students at the Portuguese academy would have used a simplified mariner's astrolabe, a graduated circle with an alidade, or rotating eyepiece, for lining up with a heavenly body, such as the North Star. What then? It is easiest to explain by taking it to extremes. The North Star is so-called because it is (almost) at the Celestial North Pole, directly above Earth's North Pole. When its altitude is 90°, i.e. straight up, your latitude is 90° N—at the North Pole. When the star's altitude is 0°, in other words it is on the horizon (and in actual fact just out of view), your latitude is also 0° and you are sailing on the equator. It was not quite as straightforward for the largest and brightest object in the sky—and the only one visible by day—the Sun, but by the fifteenth century, astronomers were steadily compiling almanac tables that gave the latitude for every conceivable solar altitude for every day of the year.

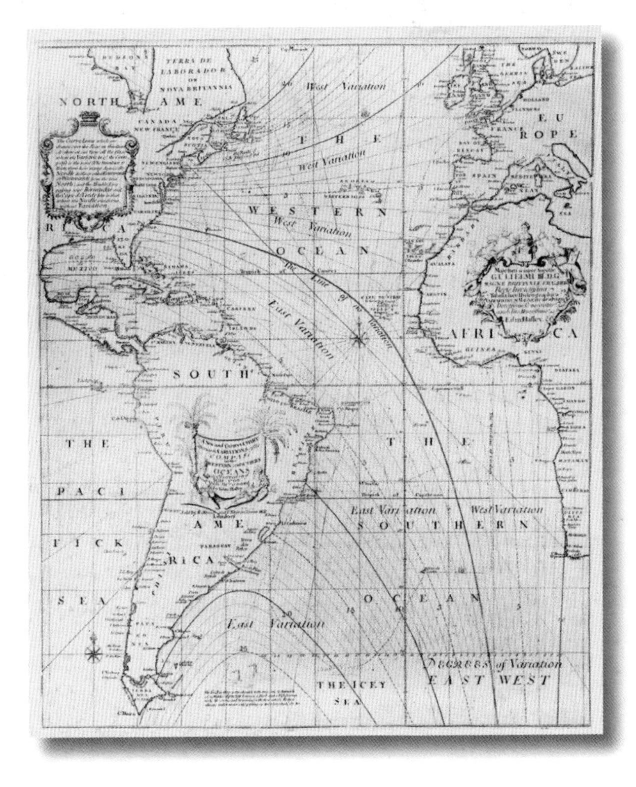

Edmond Halley made a chart of the Atlantic Ocean during a voyage in the late 1690s. The lines on the chart are isogonic: they show geographical variations in compass readings (the amount the horizontal alignment of the compass needle differs from geographical North) as a way for sailors to determine their position.

A sixth of a circle

Accuracy was crucial; a few degrees out would result in being hundreds of miles out of position. The full-circle astrolabe was also limited in that it measured the altitude relative to itself, and so had to be lined up with the horizon first—no mean feat on a rolling deck in high seas. A series of smaller, handier devices gradually evolved, finally resulting in the sextant. This had a system of mirrors that allowed the user to see the Sun (or other object) and horizon at the same time, lining them up gave the relative angle. Almanacs listed the maximum solar altitudes meaning measurements had to be performed at noon, when the sun was highest in the sky. This crucial time of day would soon become a key part of finding longitude as well as latitude.

33 The Longitude

CELESTIAL NAVIGATION WAS A HEAVEN-SENT MEANS OF ESTABLISHING POSITIONS NORTH OR SOUTH, but pinpointing the location east or west was not so simple. In the 18th century, the British government offered a huge prize to whoever could solve this crucial problem of longitude.

The Earth rotates from west to east, making a complete turn every 24 hours (give or take a minute or two), and this motion is what creates the apparent movement of the Sun through the sky each day as well as the sweep of the stars across the celestial sphere by night. To calculate latitude, navigators measured the maximum altitude of a heavenly body. This was normally the Sun, which rises in the sky until noon and then begins its journey back to the horizon for sunset. That noon-day altitude is constant wherever you are on a line of latitude, from Cape Cod to Inner Mongolia. However, noon at these locations does not happen at the same time; when it is noon in Massachusetts it is already dark in Mongolia.

Taking their time

Astronomers knew that Earth turned 15° every hour, so a location 15° west of another would have noon an hour later. An hour earlier, and you are to the east. Surely this was the solution to the longitude problem? Compare local noon with the time at a meridian of your choice—Paris or Greenwich—to calculate your position to the east or west. The problem was clock technology was not up to the job, like loading a supercomputer onto a rocket, too expensive, very bulky, and likely to go badly wrong on a long journey.

John Harrison's Sea Watch design used a spring instead of a pendulum. H5, above, was the last marine chronometer he built. In an attempt to win the Longitude Problem prize money, in 1772 Harrison sent H5 to King George III to show how accurate it was, and hopefully win his support. The following year Harrison was rewarded with £8,750—equivalent to £800,000 ($1.25 million) today.

A MADDENING PROBLEM

The *Rake's Progress* is a series of paintings and engravings made by British satirist William Hogarth in the 1730s. The central theme was the rise and fall of a young man who inherits a fortune only to squander it. Among the eight scenes Hogarth includes several jokes about 18th-century society. The final image sees the rake admitted to an insane asylum. In the background other inmates are seen tormented by insanity (plus a pair of fashionable ladies who have come to watch them for fun). Two men in the center of the image are working on the longitude problem; one searches the sky for an answer while the other contemplates a diagram of the globe. Hogarth's message is clear: the Longitude Problem drives you mad.

Guess work meets clockwork

Sailors could only estimate longitude by dead reckoning. They measured speed by running out a rope and counting the number of knots that passed in 30 seconds. The rope's knots were spaced 14 meters apart (47 feet), and a speed of 1 knot was equivalent to one nautical mile an hour (1.85hm/h). A nautical mile is the distance of one minute of meridian arc. A (very slow) speed of 1 knot due east or west would mean the ship traveled one degree (60 minutes) of longitude every 60 hours.

Dead reckoning was frequently disastrous. The British Board of Longitude was set up to administer the prize after a mass sinking (and 1,400 deaths) in 1707 caused by a navigational error. The Board was dominated by astronomers, who were convinced that a celestial method was the answer. The chief method involves measuring the angular distance between the Moon and another body. While this lunar-distance system does work, in the end the longitude problem was solved by technology. Clockmaker John Harrison spent 30 years convincing the board that his innovative ships' clocks, or marine chronometers, were accurate enough for use in nautical navigation, and were well within the criteria of the prize. In 1773, in the last years of his life, Harrison was finally rewarded for his work, but his clocks were seen as too expensive and he never officially won the prize.

34 The Age of Earth

In 1779 the Comte de Buffon, a French nobleman, challenged the established view that the Earth was created in 4004 BC with one of the first scientific investigations into the age of the Earth.

Comte de Buffon was the director of the Jardin du Roi, the French royal botanical gardens. His work had a huge influence on astronomy and biology alike, including Charles Darwin's work and his theory of evolution.

Georges-Louis Leclerc, Comte de Buffon, was more of a naturalist than an astronomer, but his work on natural history led him to wonder about the origins of Earth and the Solar System. Also an accomplished mathematician, he rejected the idea that Earth was created in 4004 BC, as a literal reading of dates in the Bible suggested. He theorized that Earth was formed when a comet had hit the Sun, and had steadily been losing heat—it was still clearly very hot beneath the surface. He knew that Earth's magnetic field suggested the planet was largely iron. So he reasoned that he could use the rate at which this metal cooled to estimate the age of Earth. He heated a small sphere of iron to white heat, waited for it to cool, and then extrapolated his results to find the time an Earth-sized sphere would take to cool. His answer of 75,000 years, still wrong of course, was a hint that the planet was much older than previously thought.

35 A New Planet

THE FIVE PLANETS (EXCLUDING EARTH) HAD BEEN KNOWN SINCE ANTIQUITY; THEY HAD NO ONE DISCOVERER, THEY HAD JUST ALWAYS BEEN THERE. Then in 1781 a seventh planet was seen and its discoverer had a name: William Herschel.

William Herschel pictured with his notes on Uranus—named George's Planet at this point—and its two moons, Titania and Oberon. Herschel discovered the moons in 1787. In the 1850s, William's son John named them for two Shakespearean characters. All of Uranus's moons are named after figures in English literature.

When he made the discovery, William Herschel, an immigrant from Germany, was an amateur stargazer. His day job was orchestral director in the city of Bath in western England, and he spent his free evenings in his back yard surveying the sky with a self-assembled (although highly effective) reflecting telescope. He was helped by his sister Caroline—also an occasional soprano soloist in her brother's concerts.

Herschel's 40-foot (12 meter) telescope was the largest in the world until 1840.

THE GREAT FORTY-FOOT

In the late 1780s, King George III asked William Herschel, by now the most famous astronomer in the world, to build the largest telescope ever near his castle at Windsor. The focal distance—the point where the telescope's mirror focused an image—was 40 feet, so the device was known as the Great Forty-Foot Telescope. The telescope design had just one mirror, which angled an image toward a viewing platform below the aperture.

Wandering disc

On March 13, 1781, Herschel angled his scope towards Gemini and saw a bright object that formed a distinct disc shape, although even the largest, brightest stars only ever form a point of light. Herschel initially thought he had found a comet, and tracked its movements over several months. By now other astronomers, including the Astronomer Royal Nevil Maskelyne, were watching the object, which Herschel had named *Georgium Sideris* (George's Star) for the king of his adopted homeland. (Later analysis revealed that the first person to record this body was John Flamsteed, the first Astronomer Royal, 90 years before, but he mistook it for a star.) The combined data revealed that Herschel's discovery was a seventh planet. Flattering the king paid off, and Herschel was elevated to King's Astronomer and moved (with Caroline) to live near the royal castle at Windsor.

Like father like son

The planet was named Uranus by German Johann Bode, who suggested that since Saturn was Jupiter's father, the next planet should be Saturn's dad. Bode also found the position of Uranus's orbit obeyed what became known as Bode's law: this assigned the progression 0, 3, 6, 12... to the planets (Earth got 6), then added 4 and divided by 10. Earth's result was 1 and those of the other planets were good approximations of their relative distances from the Sun. This law came in handy when the search for planet eight began.

36 Messier's Objects

Charles Messier was a comet hunter but he is now remembered for his catalog of objects that are not comets—but not stars either.

Messier might have been disappointed to see Uranus; he spent his nights looking for comets. He became so fed up wasting his time tracking objects that looked promising—any unstarlike blur or smudge—that he made a list of them, which he published in a catalog for fellow comet enthusiasts in 1781. The first edition had 45 objects—M1 was that pesky Crab Nebula. The document later became known as the Messier Catalog and was completed with 110 objects, which range from nebulae to galaxies and star clusters.

Far from being objects to avoid, Messier objects are greatly enjoyed by modern stargazers—albeit with better telescopes than Monsieur Messier.

37 Standard Candles

Bright but distant stars appear less radiant than dim, near ones, and so gauging the distances between stars was impossible until a teenage stargazer found a yardstick for measuring the Universe.

John Goodricke made a study of variables—stars that changed in brightness. In 1784, still only 19, he discovered one called Delta Cephei, named after its constellation, Cepheus. (Goodricke's hobby in the cold Yorkshire night reportedly killed him; he died of pneumonia in 1786.)

Forward winding to 1912, Harvard astronomer Henrietta Leavitt found a relationship between the average brightness and the pulsation period—how long it takes to brighten and then dim—of Delta Cephei (and the many other "cepheid variables" now known). So two cepheids with the same period have the same brightness, and any apparent difference in brightness is due to the distance between them. Here at last were the so-called "standard candles" that could be used to map the Universe.

John Goodricke made many observations of the star Algol in 1783. This is a variable star (although not a cepheid variable this time) which dims every three days only to become very bright once more. Goodricke, who was a deaf-mute, suggested that this variation was because Algol was a binary system in which a dimmer but larger star eclipses a small, bright one. This work won Goodricke a Copley Medal, the highest accolade of the Royal Society of London.

38 The First Asteroid Goes Missing

FOLLOWING THE DISCOVERY OF URANUS ASTRONOMERS RACED TO FIND THE NEXT PLANET. A plan was hatched to scour the sky, dividing up the zodiac—where a planet would surely be—and give an astronomer a section each.

Not every authority on the subject thought it was inevitable that more planets would be found. The philosopher Georg Hegel insisted that seven was the maximum number of planets because that was also the number of openings in the human head! Thankfully few people took any notice, and in 1800 Franz Xaver von Zach, a Hungarian baron, took charge, organizing 24 astronomers to map the sky.

One of them was Sicilian astronomer Giuseppe Piazzi. However, while still awaiting instructions from von Zach, he discovered a faint object that moved like a planet, between Mars and Jupiter. Aware that Europe's astronomers would leap on his discovery, Piazzi wanted to be sure before he announced his discovery. However, he fell ill at a crucial moment and lost his "planet" in the Sun's glare.

Giuseppe Piazzi found what later became the asteroid Ceres (now called a dwarf planet) by double-checking the positions of every observation. Ceres was the only one that had moved out of position.

Planet detectives

A frantic search began and von Zach called in the services of a German math genius Carl Friedrich Gauss. He would later be recognized as one of the finest mathematical minds in history, dubbed the Prince of Mathematics. The 23-year-old Gauss did not disappoint. It took him three months to calculate but he gave von Zach the likely orbit of the missing planet, and the Hungarian rediscovered it on December 31,1801.

Piazzi named it Ceres, after the Roman goddess of agriculture. However, although it moved like a planet, Ceres did not look like one. It was too faint to be a planet, and von Zach's team soon found similar objects—Pallas in 1802, Juno in 1804, and Vesta in 1807. William Herschel named them asteroids, and over the nineteenth century hundreds more were found, forming a belt round the Sun.

Although he is remembered for his math, Carl Gauss's official job was as Professor of Astronomy and Director of the astronomical observatory at the University of Göttingen. He is pictured here not with a telescope but with a heliometer which he used to measure the precise shape of Earth as part of his investigation into the math of complex surfaces.

39 Fraunhofer's Lines

ISAAC NEWTON HAD COINED THE WORD *SPECTRUM* AND PROPOSED ITS SEVEN COLORS IN THE SEVENTEENTH CENTURY. At the start of the nineteenth, the spectrum of light coming from stars provided the first evidence that these points of light came from objects that were made of the same kinds of materials as found on Earth.

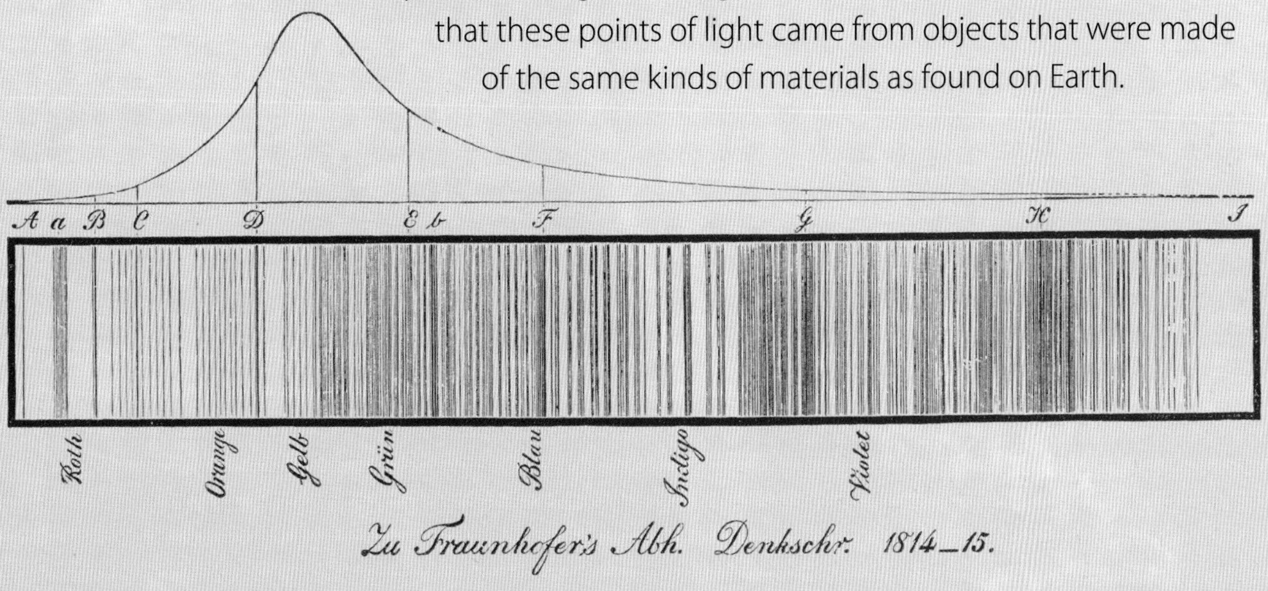

Fraunhofer's hand-drawn record shows the 570 dark lines present in the spectrum of light coming from the Sun.

As others had done before him, Newton used a prism to split white light into its rainbow of colors. The glass prism refracted the light, making its beam bend as it passed from the air into the glass and out again. This process also divides white light—supplied as a shaft of sunlight entering a darkened room—into its constituent colors because each color is refracted a different amount so they diverge into a rainbow.

More than 100 years later, technology had advanced enough to combine the prism and telescope to make a spectroscope. The first was put together by German optician Joseph von Fraunhofer in 1814, whose lenses were of outstanding quality, good enough to correct any chromatic aberration which would interfere with the results.

Missing colors

Von Fraunhofer directed his spectrometer at the Sun, focusing its light through a prism so he could see its constituent colors. However, unlike Newton he was able to observe the spectrum through a magnifying eyepiece, and saw that the rainbow was filled with hundreds of dark lines. It was as if certain colors were missing.

Fellow Germans Robert Bunsen and Gustav Kirchhoff explained the meaning of the Fraunhofer lines, which were also seen in light from stars 45 years later. This pair were chemists who found that an element could be identified by a unique spectrum of light that it emitted and absorbed. The dark lines in the sunlight corresponded to the spectra of certain elements, such as sodium, and this showed that these substances were present in the solar atmosphere. Their atoms were absorbing specific colors, leaving an absence of light that showed up as Fraunhofer's dark lines.

40 The Coriolis Effect

WITH STEAMSHIPS STILL DECADES AWAY, TRADING EMPIRES RELIED ON OCEANIC SHIPPING ROUTES FOLLOWING PREDICTABLE WINDS THAT CURVED ACROSS EARTH'S SURFACE. Sailors no doubt wondered why they rarely went in straight lines, and in 1835, they got an answer.

Objects are deflected to the right in the Northern Hemisphere and to the left in the southern half of the world. The magnitude of the deflection depends on the latitude. The surface at lower latitudes (i.e. nearer the equator) is moving faster than nearer the poles (which are effectively motionless). So the Coriolis effect is strongest at the equator.

In 1651, Giovanni Battista Riccioli suggested that Earth was not moving at all. If it were, cannonballs would swing off course as the planet moved beneath. Seventeenth-century artillery was not powerful enough to show the deflection, but by the 1830s guns and physics had advanced enough for the effect to be apparent. It was then named for French mathematician Gustave-Gaspard Coriolis. The Coriolis effect is only observed on the surface of rotating bodies such as Earth. Since the surface of the rotating object is moving, an object traveling in a straight line above the surface will trace a curved line across the surface of the body.

Coriolis suggested an extension to Newtonian laws of motion that helped calculate how bodies—the air molecules in wind or a cannonball—move against a rotating frame of reference. Coriolis showed that the apparent deflection of a flying object could be viewed as being caused by a virtual force. Although no force is actually being applied—it just looks that way—Coriolis devised a way of calculating an apparent one, so the movements of flying objects could be accurately predicted. The Coriolis force became useful for tracing ocean currents and winds (first on our planet and then on others), to track the movements of sun spots, and eventually to plan rocket launches.

The Coriolis effect is behind the claim that sinks empty clockwise in Down Under, but counterclockwise in the Northern Hemisphere. Despite this rumor, the Coriolis force does not manifest in such small systems. However it does impact on something as big as a hurricane. These monster storms are generally deflected counterclockwise north of the Equator, and clockwise in the south.

41 Stellar Parallax

THE STRONGEST ARGUMENT LEVELED AGAINST A SOLAR SYSTEM WITH THE SUN AT ITS CENTER, CIRCLED BY AN EARTH IN MOTION, WAS THAT THE STARS STAYED FIXED. From the standpoint of a moving Earth, astronomers would expect to see the stars shift in position relative to each other as the planet swung past them, a phenomenon called parallax.

Parallax is how our brains perceive distance and depth, and the effect had been used to estimate the relative distances of the Moon, comets, and planets. But when it came to the stars, they appeared (at least through primitive observation equipment) stubbornly fixed in place. Tycho Brahe had used this as evidence that Earth was not moving. To him the alternative—that the stars were so far away that the distance was as unmeasurable as infinity—was out of the question. It took until 1838 for telescope technology to become precise enough for stellar parallax, a shift in star position, to be measured by Friedrich Bessel. Bessel then used his data to show that stars were an unimaginable distance away.

Minute measurements

Using a thoroughly modern example, imagine being in a car or train as it approaches overhead cables carried on pylons running perpendicular across your path. As you approach, the pylons appear to move at different speeds, despite being in a line. The one nearest to the road or rails whizzes up toward you and flashes past, while the most distant one on the horizon appears to have barely moved at all. The parallax is the apparent motion, measured as an angle, and it can be used to calculate the distance between the observer and the object using some complex geometry. The results are expressed in terms of the astronomical unit (AU) which is the distance from Earth to the Sun. (A parsec is another astronomical distance defined as the distance to an object with a parallax of one arcsecond—a 3,600th of a degree. One parsec is about 206265 AU.)

The moon and planets are akin to the nearer pylons, while the stars are the farthest one, and any apparent motion is so tiny it alluded astronomers until Bessel used a heliometer to show that 61 Cygni (a star in Cygnus) had a parallax of 0.314 arcseconds, the first such measurement ever made. This tiny movement showed that the star was around half a million times further away than the Sun. That result translates to 10.4 light-years in modern terms, which is about 9 per cent out from the accepted value today, a remarkable achievement with hand-guided measurement.

Bessel used a heliometer built by Joseph von Fraunhofer. Its primary use was to measure the size of the Sun but it was repurposed for measuring parallax. Its lenses split an image into two and then the operator used very fine adjustments to transpose one on the other. This process could show up minute changes in the position of the object.

42 Leviathan

UPON INHERITING AN IRISH CASTLE—AND AN EARLDOM—WILLIAM PARSONS, THE 3RD EARL OF ROSSE, decided to make his seventeenth-century feudal estate the location of the world's most advanced astronomical observatory, including the largest telescope yet constructed.

Parson's monster telescope was nicknamed Leviathan and it was so huge that it had to be slung between two sturdy brick walls, making it something of an astronomical castle in its own right. Parsons had been developing techniques to make ever larger parabolic mirrors for Newtonian telescopes for several years. He cast them in several sections from speculum—the same alloy used by Newton—and built steam-powered machines for grinding and polishing them.

When completed in 1845, Leviathan had a 183-cm (72-inch) mirror, which weighed 3 tons by itself. It was fitted into another 8 tons worth of tube, constructed of wooden planks. Winches could raise and lower the massive device, but its azimuth (lateral view)—along a toothed track—was limited to about 60 degrees. Parsons used Leviathan to look again at nebulae and Messier objects. His greatest discovery was determining that many of these smudges in the sky were spiral-shaped objects filled with stars, the first view of what were later revealed to be galaxies far beyond our own Milky Way.

Looking through Leviathan was by no means easy. The eyepiece was on the side of the aperture, dozens of meters above the ground and so the observer had to climb into a somewhat precarious arcing cage that tracked the movement of the mighty apparatus.

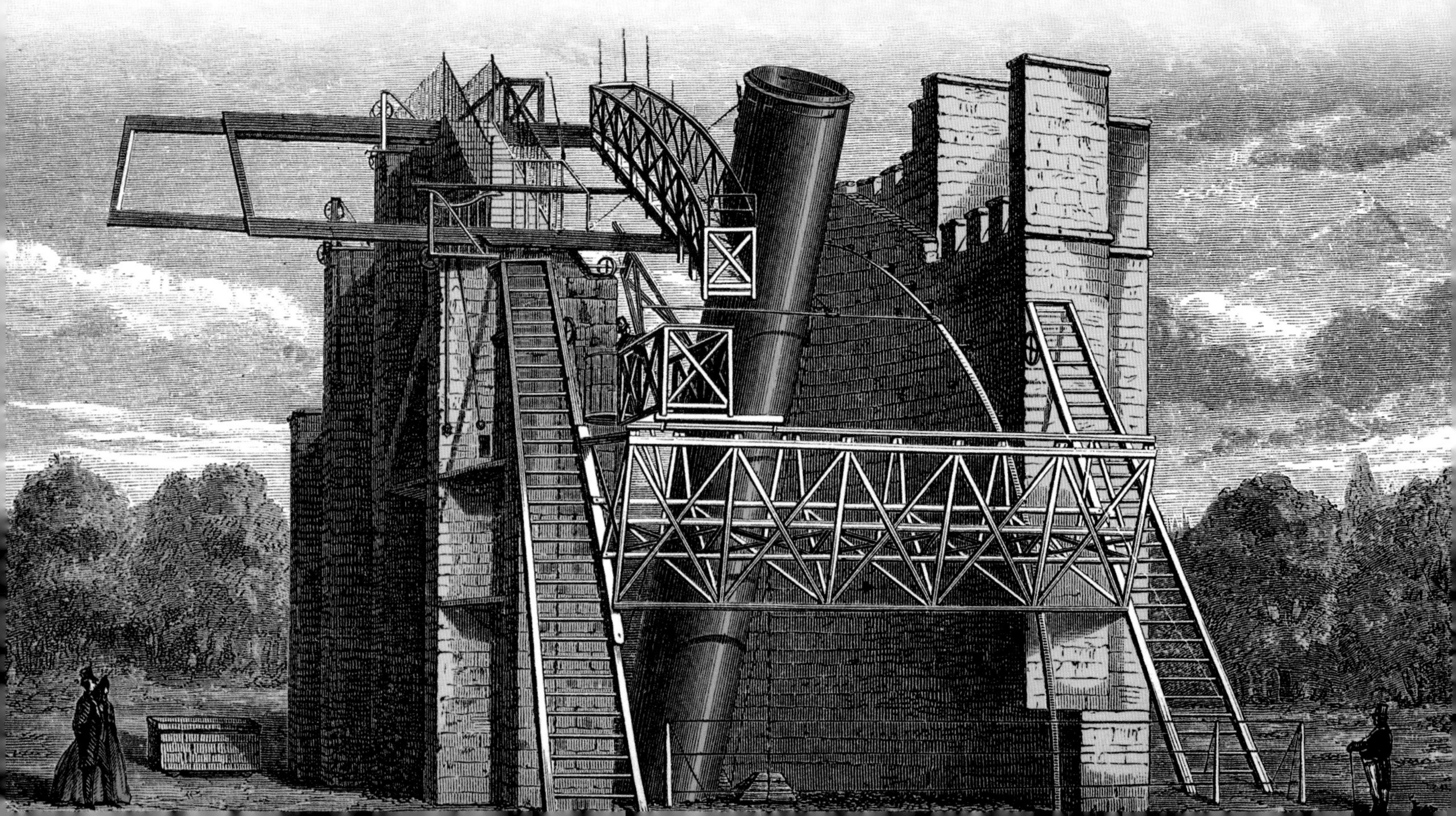

43 Neptune by Numbers

THE PLANET NEPTUNE IS THE ONLY ONE THAT IS IMPOSSIBLE TO SEE WITH THE NAKED EYE, so it is perhaps fitting that the first person to record it was telescope pioneer Galileo. However, it took math not astronomy to confirm that this was the eighth planet.

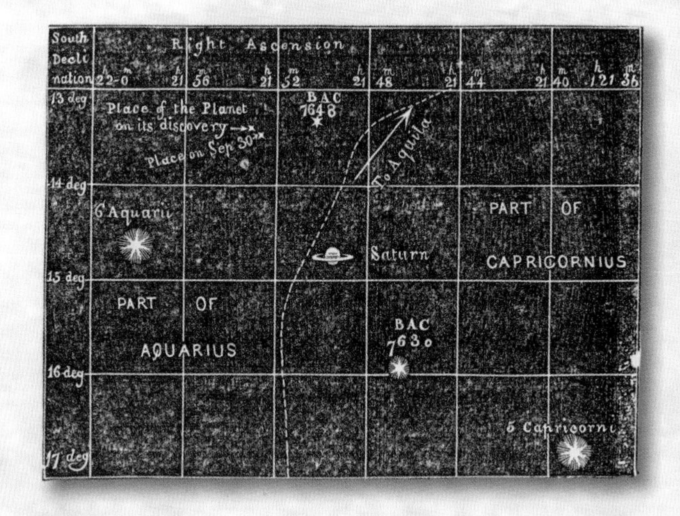

A diagram from 1846 showing the location of the new planet. It was eventually named Neptune, for the Roman god of the sea, because it has a blueish light reminiscent of the ocean.

Galileo saw Neptune way back in 1612 but he was unable to observe its movement because the planet had just started a period of retrograde motion. This is the precise moment in its orbit that Neptune appears to stop and then head back where it came from—not a real motion but an artifact of observing it from an Earth also in motion. Neptune's total orbit takes 164 years, and the chances of Galileo's telescope falling upon it at the moment it appeared to be a fixed star are astronomical, but Neptune disappeared from the records for the next two centuries.

Uranus off course

The seventh planet, Uranus, had likewise been observed by several unsuspecting astronomers prior to its eventual recognition in the 1780s. So there was already plenty of data available to compute its orbit very accurately. However, subsequent observations showed that Uranus was taking a different course. By the 1840s the prevailing theory had become that the gravity of another as yet undiscovered world, positioned further out, was pulling Uranus out of its ordained route.

Calculating the motion of three or more objects under the influence of each other's gravity had proved a fiendish mathematical problem that had defied all attempts to formalize into a general law—and does so to this day. However, the biggest and brightest math brains took up the challenge of computing a position for this mysterious new planet.

Math competition

The top mathematicians knew what was at stake and raced to find the answer. They assumed that the distance of the new planet from the Sun would approximate the next figure in Bode's law and calculated on that basis. In 1846, John Adams in Oxford completed the calculations, but it was his rival Urbain Le Verrier of the Paris Observatory who claimed the glory: Le Verrier alerted Johann Galle in Berlin, who recorded Neptune within hours of receiving Le Verrier's letter.

German Johann Galle (below) and his assistant Heinrich Louis d'Arrest got to see Neptune first because Le Verrier could not find a French astronomer interested in looking for it.

PREDICTING VULCAN

Several years after discovering Neptune in 1846, Le Verrier suggested that the solar system contained a ninth planet, this time next to the Sun. Le Verrier's argument was that a small planet, that he named Vulcan, lay between the Sun and Mercury, perturbing the latter planet's orbit. The Frenchman predicted that Vulcan orbited the Sun in just 19 days. For the next 50 years, he searched for Vulcan—in vain, despite several mistaken sightings. In 1916, Albert Einstein used his theory of relativity to explain Mercury's orbital anomalies. Vulcan did not exist.

44 Light Fantastic

UPON DISCOVERING THE GREAT DISTANCES BETWEEN THE STARS, FRIEDRICH BESSEL SUGGESTED THE LIGHT-YEAR as a unit for measuring the Universe. The light-year is the distance light travels in a year, and to know that requires an accurate speed of light.

Bessel had based his figure for the distance to the star 61 Cygni on a speed of light calculated by James Bradley in 1725. Bradley, like Rømer before, had measured the speed using observational techniques. His suggested speed was pretty accurate, but a little fast—he said that light takes eight minutes and 12 seconds to reach Earth from the Sun, which is just six seconds too quick. Nevertheless, this small error stretched over a light-year adds up to a considerable extra distance.

By 1849 Frenchman Hippolyte Fizeau measured the speed in a laboratory. He shone a light onto a mirror 8.6 km (5.1 miles) away, passing it through the gaps between the teeth of a spinning cog on the way. The cog's teeth never turned fast enough to block the light reaching the mirror. However, at a certain speed the light reflecting back did dim as a returning ray was blocked by a cog tooth swinging in front. Fizeau calculated light speed from the distance traveled and the time it took for the beam to shine through the cog and then be blocked by it on the way back. His answer was 313,300 km/s (194.675 miles/s); about four per cent out. In 1862, Léon Foucault improved the apparatus to get a speed of 299,796 km/s (186,284 miles/s)—just 4 km/s (2.5 miles/s) out!

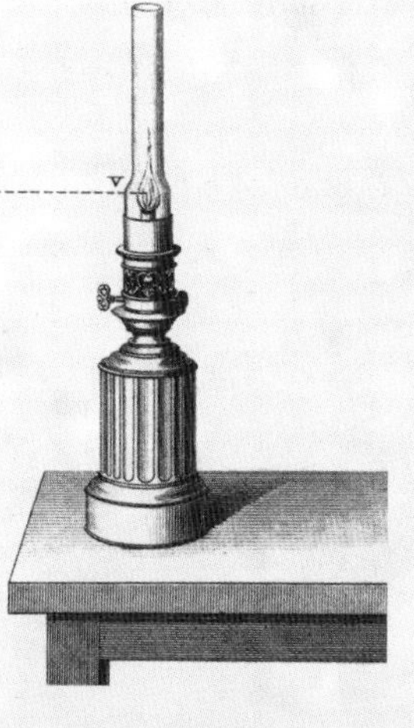

A diagram of Fizeau's original apparatus—with the long distance taken out. Lamplight was reflected through a primary telescope, past the spinning cog fitted near the eyepiece. The light was focused by a second telescope and then reflected straight back into the first.

45 Foucault's Pendulum

One of the central tenets of the Copernican solar system was that Earth rotated around the Sun once a day, creating the illusion of motion in the heavens. Despite all the clues it had proven to be impossible to observe this first hand while the observer was confined to the surface of the planet. That is until Léon Foucault erected a large pendulum in Paris and set it swinging. Here, at last, was proof the world turned.

The motion of a pendulum obeys certain laws. Foucault's genius was to show that any behavior outside of those rules must be down to the movement of something else—if it was not the pendulum moving then it was the whole Earth instead.

Foucault's Pendulum was demonstrated all over the world, including in London as shown above. The original was swung in the Pantheon in Paris, where a replica is still in situ.

In the swing of it

Legend has it that Galileo discovered the principle of the pendulum as a student in 1582 while watching a heavy lamp swinging from the ceiling of the cathedral of Pisa, Italy. He timed the swing of the lamp, using his own pulse as a clock, and realized that, though the size, or amplitude, of the swing diminished each time, each complete back and forth oscillation of the lamp always took the same amount of time. Galileo later showed that the period of oscillation of a simple pendulum is proportional to the square root of its length. Changing the mass of the pendulum bob has no effect—a heavier bob oscillates with the same frequency as a lighter one. Isaac Newton later explained that inertia—a resistance to change in motion—would keep the pendulum swinging in one plane.

Apparent deflection

In 1851, Foucault set up a heavy pendulum in Paris. As it swung it traced its path in a circle of sand on the floor with a fine point on its underside. At first the bob swung in a fixed plane, but after many hours the path of the pendulum gradually appeared to rotate clockwise. Nothing had been allowed to interfere with the pendulum's swing so this apparent change in direction was due to the sand bed—and Earth—moving beneath the pendulum. After 24 hours, the pendulum returned to its original plane—Earth had made one complete rotation just as Copernicus had said.

46 The Sunspot Cycle

THE FIRST LESSON OF ASTRONOMY IS NEVER LOOK AT THE SUN DIRECTLY THROUGH A TELESCOPE OR ANY FORM OF LENS. The naked eye can protect itself from bright light but a intensified, focused beam will damage the retina permanently. So, studying the Sun required a novel approach.

The best way to observe the sun is to project an image of it onto a wall or other pale, blank surface. Some ancient astronomers did this using a camera obscura, a kind of room-sized pinhole camera, while others may have taken a look through smoked glass filters. Whatever the method, the most obvious surface feature of the Sun is the sunspot, which was first recorded in the fourth century BC by Chinese astronomers.

There were occasional reports of large dark patches on the sun over the ensuing centuries, most often interpreted as Mercury in transit across the solar disc. However, in 1612, when Galileo projected the Sun through his telescope, he correctly identified them as spots on the star's surface, and tracked their appearances and movements.

A detailed sketch of the structure of a sunspot was made by Italian Jesuit astronomer Father Pietro Angelo Secchi in 1873. It shows the central umbra surrounded by the penumbra. These features only look dark because they are surrounded by even hotter material. If the spot was alone in space it would radiate bright light—the average sunspot is twice the size of Earth!

THE LITTLE ICE AGE

In the early twentieth century, the English astronomer Edward Maunder used historical records to trace solar activity back into the seventeenth century. He found that between 1645 and 1715 there were almost no sunspots at all. This period, known as the Maunder Minimum, appears to have had an impact on the world's climate. From the middle of the seventeenth century to the middle of the next century, the weather was distinctly colder. It is now known as the Little Ice Age in Europe. The River Thames froze almost every winter, and Frost Fairs (below) were held on the thick ice.

A pattern develops

William Herschel subscribed to the view that sunspots were gaps in the furnace of flaming clouds that covered the surface of the Sun, showing a cooler, darker surface below. He also, not unusually for an eighteenth-century astronomer, suggested that people lived in beneath the fiery atmosphere. Only one of his intuitions was correct though: sunspots are cooler than their surroundings but are still unimaginably hot.

The first person to suggest that sunspots followed a cycle was the German Heinrich Schwabe. He made an extensive study of them over 28 years from 1826 to 1843. He began this survey because he believed in reports that a small and fast-moving planet, generally referred to as Vulcan for the Roman god of fire, existed inside the orbit of Mercury. Schwabe's theory was that this planet would be all to easy to lose among the sunspots, but its motion would give it away. Try

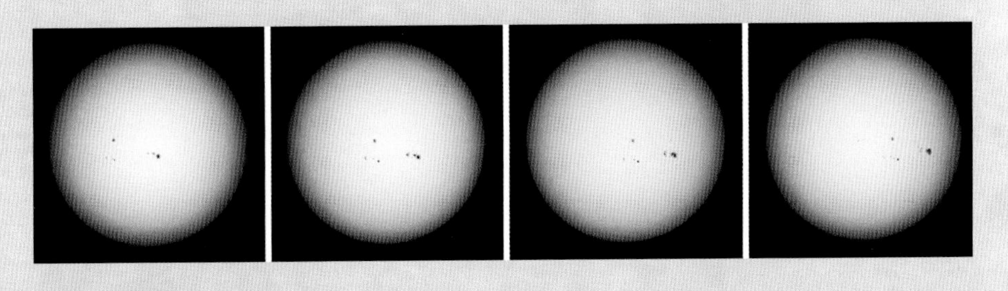

Time-lapsed images show sunspots drifting across the solar surface over a period of four days, analogous to Earth's weather systems. Since the Sun rotates, each sunspot's path is subject to the Coriolis effect.

as he might, all he saw were sunspots. However his data revealed that the number of spots rose and fell in a cycle roughly 11 years long.

Once Schwabe published his data in 1851, Swiss astronomer Rudolf Wolf combined it with the records of others to track sunspot behavior all the way back to the 1740s. Peaks in sunspot numbers were roughly 11 years apart and in between there were periods with few if any sunspots.

In 1908, American George Hale revealed that sunspots were twists in the Sun's magnetic field, and that intense knots of magnetism pushed away the rising heat, making the spot cooler. This cycle of magnetic field "maxima" and "minima" is due to the field becoming increasingly tangled as the star rotates before eventually resetting itself.

47 Helium: Sun Gas

ANOTHER WAY TO STUDY THE SUN WAS TO OBSERVE FULL SOLAR ECLIPSES, WHERE THE GLARE OF THE STAR is hidden by shadow and the wispy corona of hot gases that surrounds it comes into view.

In the total eclipse of 1868, spectroscopes turned to the corona, which was found to be full of dark lines. By now Gustav Kirchhoff (following work with Robert Bunsen) had codified the science of spectroscopy with three laws: the first law states that hot solids produce a full spectrum of colors (seen as white light); number two says that hot gas (like a flame) glows with a specific set of colors (known as its emission spectrum); and lastly cold gas absorbs specific colors from white light leaving dark lines in the full spectrum (termed the absorption spectrum). These rules allowed astronomers to analyze the makeup of stars, nebulae, and interstellar dust.

Helium is the by-product of solar fusion, the process that powers the Sun. Its emission spectra, seen below, contains the distinctive yellow that led to the gas's discovery.

The Sun's corona was hot enough for its gaseous content to present itself as emission spectra. The 1868 eclipse offered up a distinct yellow line in coronal light that was seen by Pierre Jansen. Norman Lockyer tried to reproduce the color with known elements until 1870, whereupon he declared that the corona contained a new, unearthly gas, which he named helium for *helios*, "sun" in Greek.

48 Martian Canals

AS TELESCOPES IMPROVED, THE SURFACE OF MARS GRADUALLY CAME INTO FOCUS. In 1877, Earth passed very close to the red planet, affording the best view for a generation. But the fallout from a mistranslated map of Mars made that year still resonates today.

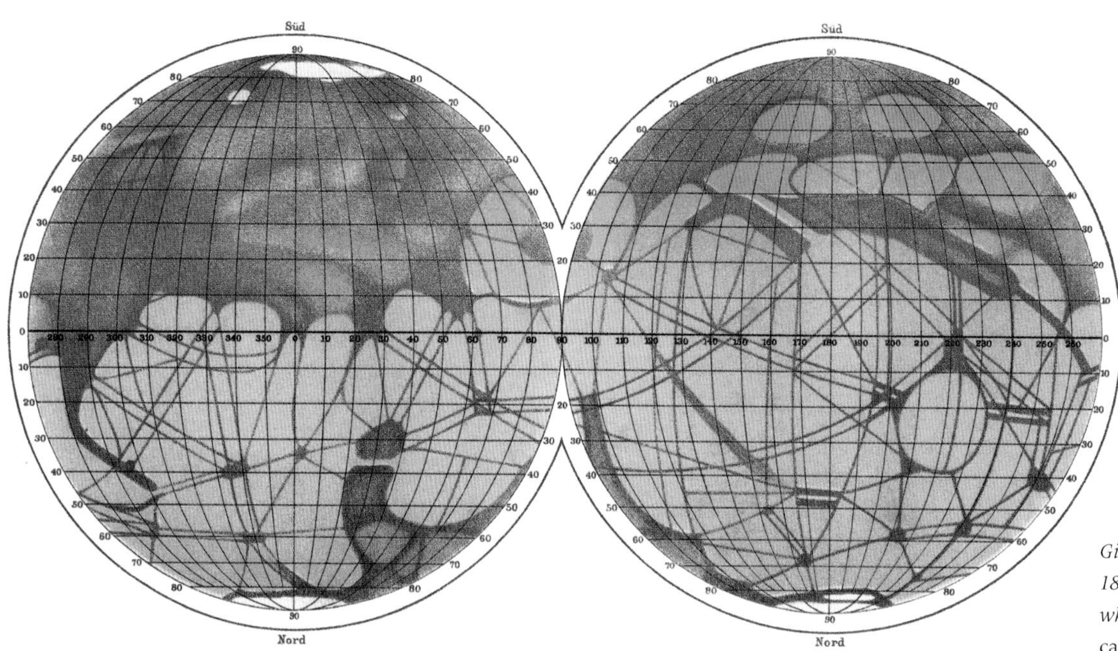

Giovanni Schiaparelli's 1877 map of Mars showing what he described as its canali. *More recent surveys of the surface have revealed that there are many erosional features, probably formed by a shallow water ocean that covered the now-dry planet in its early history.*

Schiaparelli chose to map Mars during a "great opposition" when the Earth is between the Sun and Mars and is also at its closest point to its red neighbor.

Cassini and Huygens had reported that Mars had ice caps at the poles and dark regions on its surface. William Herschel used his huge telescope to reveal that the ice caps shrunk and grew each year just as on Earth. The dark regions also grew in size—Herschel suggested that meltwater from the poles was flooding the surface. Others thought that the color changes, which developed over a matter of weeks, were patches of vegetation sprouting in the Martian spring. (We know now that the dark areas are bare rock exposed by huge storms blowing away paler dust.)

In 1877 the Italian Giovanni Schiaparelli saw what could be rivers connecting the dark "oceans." He marked them as *canali*, meaning channels, on a detailed map, but that was translated as *canals* in English—artificial waterways. This seeded the idea that Mars was home to aliens who appeared industrious, and being Martian were perhaps belligerent, too—H.G. Wells certainly thought so with his *The War of the Worlds* novel. Many amateur astronomers began to report seeing canals, including Percival Lowell, a wealthy American businessman. He built an observatory at Flagstaff, Arizona, with the express intention of searching for signs of Martian life. The Lowell Observatory found none but it did stumble across Pluto in 1930.

49 Standardizing Time

UNTIL THE MID-NINETEENTH CENTURY, TIME WAS A DISTINCTLY LOCAL AFFAIR. Midday was when the Sun passed overhead, wherever you were. However, that played havoc with the timetables of the new railroads—and nothing, not even astronomy, could stand in the way of industrial progress.

In 1883, the United States adopted its four time zones. This slightly later map also shows the Atlantic Standard Time Zone used by the maritime provinces of Canada.

Any navigator would have known that for every degree that a train traveled, its local time changed by four minutes. This first became truly apparent on the Great Western Railway between London and the western port of Bristol, a city that measured time ten minutes behind the capital, causing understandable confusion. The solution was to use Railway Time, a standard time based on the average day length measured at Greenwich—Greenwich Mean Time (GMT).

There was considerable opposition. People were proud of their local time, while railway time was seen as an insidious technology to be treated with suspicion, much like modern attitudes to the inevitable homogeneity of multinational corporations. Nevertheless GMT won out.

World standard

By the 1880s the problem had gone global, due not only to faster transport systems, but also to the telegraph network that connected far flung places in an instant. In 1884, Canadian astronomer Sandford Fleming called an international conference in Washington, D.C., with the aim of dividing the globe into time zones based on a standard time. The British Empire and United States already worked to GMT and used nautical charts based on that meridian so that was the obvious choice. Countries were left to choose which time zone they wanted to use. France objected to GMT and continued using Paris as the "prime" meridian until 1911.

The Exchange in Bristol, England, still uses a clock with two minute hands—one shows local time, ten minutes behind GMT.

50 Space Travel

LEGEND HAS IT THAT THE FIRST SPACEMAN WAS WAN-HU, A CHINESE ADVENTURER who, in the sixteenth century, attached 47 rockets to a chair. Once lit, the rockets produced a cloud of smoke—Wan-Hu was never seen again!

A sketch from one of Tsiolkovsky's notebooks which are on display in a museum at Kaluga, Tsiolkovsky's hometown south of Moscow.

THE BRICK MOON

Space travel appeared first in fiction, such as Jules Verne's 1865 novel *From the Earth to the Moon.* This paid little attention to the laws of physics, unlike *The Brick Moon* by Edward Everett Hale in 1869. Although still rather fanciful, this short story outlines the events when a brick sphere—intended to be sent into orbit as an easily visible navigational aid—is launched by accident with people aboard. This was the first record of the concept of an artificial satellite and space station.

A modern-day reconstruction of this event showed that more than likely Wan-Hu's craft would have exploded on the launch pad, and rocket technology was focused on the weapons sector for most of its history. The "rockets' red glare" lyric in the U.S. national anthem refers to a bombardment of American harbors with Congreve rockets fired from British warships in 1812. These rockets, effectively giant fireworks with metal cases, were in turn inspired by Indian war rockets.

Space visionary

Russian teacher Konstantin Tsiolkovsky was the first person to suggest using rockets to fly into space. He determined that the Earth's escape velocity was 8 km/s (5 miles/s). Using a formula now known as the Tsiolkovsky rocket equation, he showed that this speed could be achieved using the thrust of a rocket powered by super-chilled liquid hydrogen and liquid oxygen—the fuels used by the largest rockets today.

Tsiolkovsky's book of 1903 also predicted many aspects of space travel, including double hulls for protection against meteor strikes and the health problems of weightlessness. He later designed a multistage rocket, which he described as a "rocket train," which dropped empty sections to reduce weight as it flew higher. In 1911, Tsiolkovsky proposed a crewed spacecraft, in which the passenger lay face-upward on the floor near the top of the rocket so as to better withstand the crushing gravitational forces during the flight

Konstantin Tsiolkovsky with models of his rocket designs in 1919. Although he never built a rocket, Tsiolkovsky's work was highly influential in the development of Soviet rocket and space technology.

51 The Tilt of the Earth's Axis

THE CELESTIAL SPHERE OF STARS, THE THREE-DIMENSIONAL MAP OF THE SKY USED BY ASTRONOMERS, HAS EARTH AT ITS CENTER. However, the precise position of Earth in relation to the Sun is always shifting slightly, and needs constant monitoring to insure that the star charts stay accurate.

Simon Newcomb was not afraid of making predictions. In 1888 he said, "We are probably nearing the limit of all we can know about astronomy." In 1903 he also declared that flying machines were impossible with the materials currently known to science—within months the Wright Brothers proved him wrong.

Since Hipparchus, astronomers had known that Earth's axis was prone to wobble extremely slowly, a process called precession that changes the angle between the axis and the plane of the ecliptic. The latter is the imaginary plane in space in which Earth moves around the Sun. Every heavenly body is located in the sky by its position relative to where the celestial equator (which matches Earth's) crosses the ecliptic.

Astronomers had to make frequent updates of the relative positions of heavenly bodies to keep up with precession. In 1895 Canadian-American Simon Newcomb presented a mathematical means to predict the relative positions of the Earth and Moon in his *Tables of the Sun*. These tables were used until 1984 when NASA introduced an improved version that was based on the latest measurement of the Solar System using space-age accuracy.

SEASONAL EFFECTS

The mismatch between the axis and the ecliptic is what creates Earth's seasons. In summer, the Northern Hemisphere of the planet is tilting toward the Sun. That makes the Sun travel higher into the sky, meaning the days are longer (and therefore warmer). The south meanwhile is in winter with a lower Sun, and shorter, cooler days. Six months later, Earth has moved to the other side of the Sun. Now the north is tilted away, and it is the Southern Hemisphere's turn to enjoy better weather.

52 The Speed Limit of the Universe

"WHAT WOULD YOU SEE IF YOU SAT ON A BEAM OF LIGHT?" THAT IS THE QUESTION THAT ALBERT EINSTEIN IS REPUTED TO HAVE ASKED HIMSELF WHILE STILL A TEENAGER. The answer was a paradigm shift as huge as the discoveries of Copernicus, a means of showing how energy, matter, space, and time are harnessed together to create the Universe.

So what is the answer? It is time to dispense with your intuitions, since much of Einstein's 1905 theory of special relativity is almost nonsensical. You might think that since a beam of light travels at the speed of light, when looking backwards as you whizz along on your photon, no other beam of light can catch up with you from its source—a star or whatever—and so your eyes see nothing at all. Not so said Einstein. And looking forwards, the light coming the other way (also presumably at the speed of light) must be traveling at twice the speed of light relative to you the observer. Impossible said Einstein. Everything would just look normal. If you measured it, light arriving from all directions would be traveling at the same speed, irrespective of your speed relative to the sources.

Albert Einstein became the archetypal genius scientist, lending his wild hair and middle-European accent to countless eccentric professors and unhinged inventors in children's cartoons.

Aether wind

This suggestion was at odds with the theory of light that had prevailed throughout the nineteenth century. As a wave, light needed a medium to pass through. Just as sound was propagated by air, so light was thought to be carried by luminiferous ether—a universal, barely substantial background material akin to Aristotle's quintessence from 2,100 years ago. Since Earth was moving through the ether as it rotated, light beams traveling perpendicular to Earth's motion would experience "drag" leading to a tiny shift in direction as its ethereal medium moved sideways, "blowing" the light off course. The Michelson-Morley experiment of 1887 was set up to detect the effects of ether drag. There was none, thus dispelling the theory of ether once and for all and demanding a new explanation.

THE TWIN PARADOX

Traveling close to light speed also makes a mass move much more slowly through time than a stationary one does. Imagine a twin sent on a long space voyage on a super-fast starship that travels at near light speed. Our explorer would not notice any difference to time while on board—to him the clocks move at a normal speed. He left on his 25th birthday and returns to Earth on his 26th, obviously eager to have a birthday celebration with his twin sister. However, at the party, the sister's cake requires many more candles. A year traveling at great speed is equivalent to dozens staying (relatively) still down on Earth.

Spacetime

Einstein's idea was to bring all dimensions into a single cohesive spacetime and relate them to energy and mass. Mass curved spacetime and this warping manifests as the force of gravity. As a mass moves ever faster, space contracts making the object shrink in the plane in which it is moving. Its mass also increases and more energy is needed to move it faster. If it were to move at the speed of light, the object would have to become infinitely massive and require infinite energy to achieve that speed—obviously impossible. Therefore no mass can travel at the speed of light, only the massless photons that make up light itself. These changes in mass and space are imperceptible on the human scale (although they can be measured) but they ensure that light speed is constant—relative to any observer no matter what his or her own speed.

53 Cosmic Rays

IT HAD BEEN KNOWN FOR SOME TIME THAT AIR WAS SLIGHTLY CONDUCTIVE. In 1912 a daring balloon-based scientist showed how this conductivity was due to rays arriving from outer space.

Austrian physicist Victor Hess prepares to fly to 5,000 meters (5,456 yards) in 1912 to test the conductivity of air at high altitudes.

An electrically charged object either has too many or not enough electrons. On Earth, this charge is eventually lost because charged particles (ions) in the air balance out the number of electrons, taking them away or making up the shortfall as required.

Advances in atomic physics had revealed how gases in the air become charged ions when they are bombarded with high-energy rays. In 1911 Austrian physicist Victor Hess began making high-altitude balloon flights to investigate how the conductivity of the air varied with height. He carried with him charge detectors called electroscopes. These were fully charged on takeoff, so two gold foils within carried the same charge and thus repelled each other. As the electroscope lost its charge to the air, the foils sunk ever closer together. Hess found that at higher altitudes, the electroscopes lost their charge more quickly. The thin air up there was more heavily ionized by what later became known as cosmic rays—high-energy particles and radiation streaming out of exploding stars and constantly bombarding Earth's atmosphere.

54 Star Types

BY THE START OF THE TWENTIETH CENTURY ASTRONOMERS HAD A NUMBER OF WAYS OF COMPARING STARS OTHER THAN THEIR LOCATION AND A COMPARISON OF THEIR BRIGHTNESS. All these measurements created a mind-boggling cacophony of data until two astronomers helped to make sense of it all with a simple graphical representation. Within the diagram was also the first chapter of the life story of stars.

The brightness of a star is termed its magnitude. Hipparchus started the system by assigning each heavenly object to one of six magnitudes. The magnitude scale we use today was set up by Norman Pogson in 1856. He gave bright stars (but not the brightest) like Altair, a magnitude of 1. William Herschel had noticed that stars of the sixth magnitude (in the ancient Hellenic system) were 100 times brighter than those in the first, Pogson made magnitude 6 stars 100 times less bright than magnitude 1, with 2 to 5 equally spaced in between. Magnitude 7 stars are too faint for the naked eye, while the brightest objects of all have negative magnitudes: Venus is -4, a full Moon is -12.6 and the Sun is -26.7!

Every star has an apparent magnitude—how bright it appears in the sky—and an absolute magnitude, a measure of how bright it is compared to other objects. The

The comparative sizes of stars is somewhat mind-blowing. Our Sun is the orange orb. Behind it is Sirius A, a blue-white star roughly 1.7 times the size of the Sun. The red star is Proxima Centauri, our closest star after the Sun and about the size of Jupiter. The small dot is Sirius B, a white dwarf, smaller even than the planet Earth. However, if the Sun was seen at the size of Sirius B, a supergiant would easily appear as large as Sirius A in this picture. The largest star every found, VY Canis Majoris, is 2,000 times the width of the Sun!

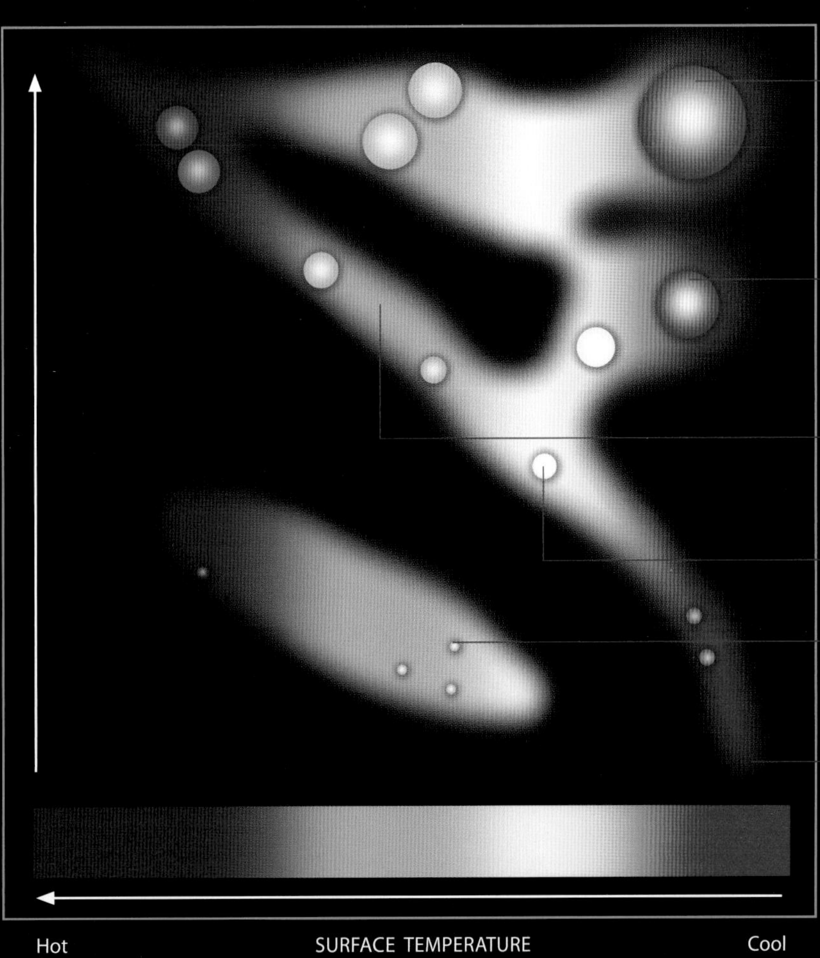

Supergiant stars form from main sequence stars that are several times larger than the Sun. They burn quickly and die young.

Giant stars are most often red. They form when smaller main sequence stars begin to run out of fuel.

Main sequence stars are young or middle aged with billions of years to burn.

Our Sun is a decidedly average yellow dwarf star.

White dwarfs are the Earth-sized hot cores of long-dead red giants.

The tail end of the main sequence is composed of brown dwarfs, balls of gas that are not quite massive enough to give out much light.

The Hertzsprung-Russell, or H-R diagram, was first put together by the Dane Ejnar Hertzprung who plotted stars' brightness against their colors in 1911. Two years later the American Henry Russell plotted brightness against surface temperature (indicated by color) to produce the version we use today.

latter can be calculated from the former once the distance is known. In addition, by studying the way binary stars orbited each other—watching how their light changed as one moved in front of the other—astronomers could calculate the mass of stars using Newtonian laws of gravity. It was found that stars came in all sizes, some being thousands of times the size of our Sun.

Hertzsprung and Russell

Spectrographic surveys of the stars revealed that they were not all made of the same elements. Astronomers began to classify stars according to the substances they found in their atmospheres. Blue stars were also hotter than red ones, and color became equated with surface temperature. By 1913 two astronomers—Ejnar Hertzsprung and Henry Russell—had independently plotted the magnitudes and temperatures on graphs. They found that the stars are not splattered at random, but most, including the Sun, formed a main sequence running from hot (blue) and bright to cool (red) and dim. The stars in this central belt were termed dwarfs—the Sun is a yellow dwarf—to differentiate them from those outliers that were clustered above the main sequence. These were the giants—bright but also generally cool. Below the main sequence were dim but hot stars called white dwarfs—not quite hot enough to be blue. Future investigations into how these different stars formed would lead to the very creation story of the Universe.

55 Bending Time and Space

IN 1796, THE UNDOUBTED FRENCH GENIUS, PIERRE SIMON LAPLACE CONSIDERED THE POSSIBILITY of an object with gravity so strong that even light could not escape from it. This *corps obscure* was just a thought experiment, but by 1916 Einstein's general theory of relativity had suggested it was real after all.

This is an illustration of an Einstein ring, where light from a distant galaxy bends around a black hole in between it and the Earth. All light follows the warps in space but its effect is most marked around a black hole.

A decade after turning physics on its head with special relativity, Einstein had developed a general theory which put his ideas into the context of the everyday Universe. It was an update to the Newtonian laws of gravitation, which although flawless for predicting the motion of baseballs, the Wright Brothers' new-fangled aircraft and the trajectory of artillery shells, could not quite explain every intricacy of planetary motion. Even tiny inaccuracies when cast across the increasingly well-understood vastness of space added up to major errors.

Straight lines curve

To fix these problems Einstein set out a Universe in which space and time were four facets of the same thing. (He eventually suggested that more dimensions were at play.) That meant that geometry of space was not quite how we perceive it. The shortest route from one point to another is always a straight line. However, in space such a straight line is curved—and may make several twists and turns. This is because mass bends space, and lines on curved surfaces follow a different set of rules. If you measured these distance along these curved lines with your hypothetically immense ruler, you would find it to be perfectly straight. But that is because your ruler is curved by the warp and weft of spacetime, too.

The amount of curvature in space depends on the amount of mass. The Sun creates more of a depression in space time than Earth, a deeper "gravity well." The Earth is pulled toward the Sun, down into its gravity well. Fortunately our planet's orbital

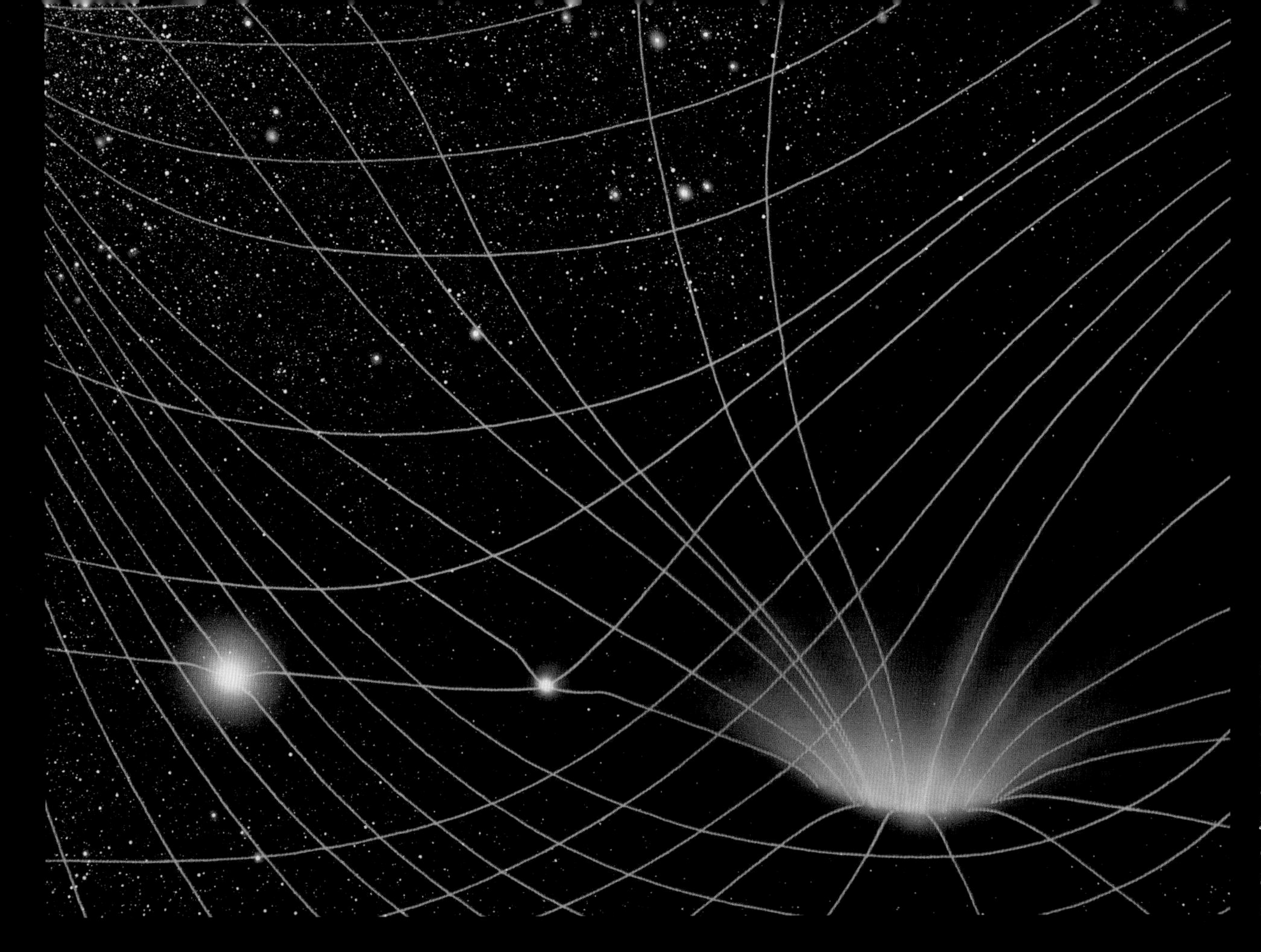

General relativity explains that all mass bends space making dips. A black hole is named after the depth and utter darkness of its gravity well. That presented a problem in finding them out in space. By definition they do not shine.

velocity is high enough so we just roll around the well and not into its less hospitable center. This is a good way of visualizing the way the force of gravity acts—imagine Newton peering into a gravity well with an apple plummeting into it.

Einstein's theory also explains some more extreme effects. It predicted that the path of light arriving from a star located near the edge of the Sun—when seen from Earth—would be curved as it passed through the Sun's gravity, making it appear out of position. Such peripheral stars were hidden in the glare of sunshine, but in 1919, Arthur Eddington measured their positions once they became visible during a solar eclipse. His results supported Einstein's theory. Relativity was reality!

Karl Schwarzschild's prediction of black holes was known as the Schwarzschild radius initially. The term "black hole" was not coined until the 1950s.

Schwarzchild's brainchild

In 1915, while Einstein was still adding the finishing touches to his general theory, the great man published some field equations that gave the relationships between energy, mass, space, and time. Karl Schwarzschild, a mathematician who was taking time off to fight the First World War, used them to calculate how small a star had to be before its escape velocity (the speed needed to escape its gravity well) reached the speed of light. Unlike Laplace, Schwarzchild knew that the speed of light was an unexceedable constant, and that this *corps obscure*—later "black hole"—would be a strange object indeed. The Schwarzchild radius is the size of the event horizon, an imaginary line in space named for the concept that anything passing it can never cross back again. We can never look inside a black hole, because nothing every comes out of one, not even information. However, 60 years later a glimpse of the interior would be revealed.

56 Islands in Space

ALL THAT TWINKLES IS NOT A STAR. GALILEO HAD SEEN THAT THE PALE HAZE OF THE MILKY WAY WAS PRODUCED BY STARS TOO NUMEROUS TO COUNT. William Herschel had mapped these stars into a flat disk, making our Solar System part of what became termed the "Galaxy." However, when astronomers began to see objects that appeared outside the galaxy the question arose: is there more to the Universe than the Milky Way?

In the early 20th century, the Dutch astronomer Jacobus Kapteyn carried out the most extensive survey of the Milky Way to date. He found it to be a flat, dish-shaped region that bulged in the center and became gradually more diffuse, as stars became more widely spaced toward the outer rim. In line with the thinking from Herschel onwards, this showed an "island universe" 60,000 light-years wide (later shown to be five times that size) and about 10,000 light-years thick.

The idea of an empty, black Universe with a single glowing cluster of stars (where Earth happened to be located) appealed to many people, but astronomers had their doubts. Even William Herschel pondered the prospect that some of the nebulae, the "little clouds" or amorphous smudges of light, were themselves distant islands within the void of space. He eventually dismissed the idea, but it lingered in the work of catalogers such as Messier, who found a wide range of nebulous objects that seemed unlikely to all have a common origin. The Leviathan of Parsonstown, the giant Irish telescope, had shown that several nebulae shared the swirling disk shape of the Milky Way. The question remained, were these objects in the Milky Way or further afield? It would take even larger telescopes to answer these questions.

GALAXIES COLLIDE

From our earthly point of view, galaxies are rather empty with great distances between neighboring stars. However, a galaxy is pulled together by the tiny gravitational forces between stars, and as a whole galaxies pull on each other. Every so often, gravity makes galaxies collide, fusing together into a single larger one. The Mice (below) are colliding galaxies named for their long tails.

Edwin Hubble used the world's most powerful telescope to provide conclusive proof that the Universe comprised more than one galaxy—it turns out to be quite a few more.

Light and distance

By 1908 more than 15,000 nebulae had been recorded. They appeared to fall into two broad groups: diffuse blobs that were located near to the Milky Way, and symmetrical disks and spirals that were harder to locate. Spectrographic analysis of the light from these nebulae revealed that members of the first group were clouds of cool gas—with the odd star within—while the second set gave out light similar to that seen from individual stars.

The action then moved to Mount Wilson, California, where in 1917 a telescope larger even than Leviathan began to look at nebulae. The Hooker Telescope had a mirror 100-inch (254-cm) wide and would rule as the king of telescopes for the next 30 years. One of the first things this telescope saw were novae—bright "new" stars appearing as if from nowhere—with some of the nebula. They were much dimmer than the novae observed in the Milky Way, but assuming that all novae were of similar magnitude, this would show that those nebulae were a million light-years away! This remained conjecture for another few years until, in 1924, Edwin Hubble at the Mount Wilson Observatory found cepheid variables in Messier objects 31 and 33 and other disc-shaped nebulae. The faint light from these markers showed once and for all that these objects were far beyond the limits of the Milky Way. They were not nebulae at all—that name is now reserved for clouds of space gas—but were galaxies, island universes just like our own.

Later research showed that galaxies actually exist in clusters. The Milky Way shares the Local Group (an inspired name, perhaps) with the Andromeda Galaxy, Magellanic Clouds and 30-odd other galaxies. In turn, the Local Group is in the Virgo Supercluster, along with 100 other galaxy clusters. The numbers are mounting up: a conservative estimate of the total number of galaxies is 125 billion!

This spiral galaxy was first identified as Messier 81, but is now named Bode's Galaxy. It is 12 million light-years away. At its heart is a black hole with a mass of 70 million Suns.

57 Rocket Robert, Space Pioneer

SOLID-FUELED ROCKETS ARE MORE THAN A MILLENNIUM OLD—THEY ARE EFFECTIVELY LARGE FIREWORKS. Space rocket pioneers believed that the extra boost of liquid fuels was needed to reach space, but rocket technology needed to catch up first.

Space travel needs an engine that can work anywhere—on the launch pad, in the high atmosphere, and in the vacuum of space. Air-breathing engines—such as external combustion steam engines and internal combustion gasoline engines—would fail as the air thinned out at high altitude. The oxygen in the air is needed for the fuel to burn and release energy.

Not so with rockets, which are slightly mischaracterized as being liquid- or solid-fueled. In fact, liquid-powered rockets carry two fuels—a propellant and an oxidant—which react violently when mixed, creating hot, rapidly expanding exhaust gases. This gas is directed out of a single nozzle, and the laws of motion do the rest: the rocket pushes the gas backwards, so the gas pushes the rocket forward—very fast. The other great advantage of liquid-fueled rockets is that they can be turned on and off, a crucial requirement for any budding space explorer.

Robert Goddard with an early liquid-fueled rocket in 1927.

Dream of space

The story goes that Robert Goddard felt the urge to fly to space while climbing a tree as a teenager. He imagined a space rocket fueled for launch in the field below him, ready for a mission to Mars. Seventeen years later, in 1926, Goddard launched the first liquid-fueled rocket, with many to follow reaching speeds just below the sound barrier. His device ran on gasoline and liquefied oxygen (kept liquid by low temperatures and high pressure). The first flight took place in the snows of New England, which helped chill the oxygen, but it lasted just a few seconds before the nozzle burned off and the device crashed into a field of cabbages. Several more designs resulted in the now classic formation of fuel tanks connected to a combustion chamber at the base of the rocket.

One of Goddard's rocket launches in 1937. By this time his "vehicles for reaching extreme altitudes" as he called them were being superseded by those of others, not least by rockets being built in Nazi Germany.

58 The Expanding Universe

REDSHIFT

The color of light is the way our brains perceive the wavelength of light. Red light has a longer wavelength than blue. When an object moves away the wavelength is stretched making its colors shift toward red. Blueshifted light has had its wavelength compressed as its source moves towards the observer.

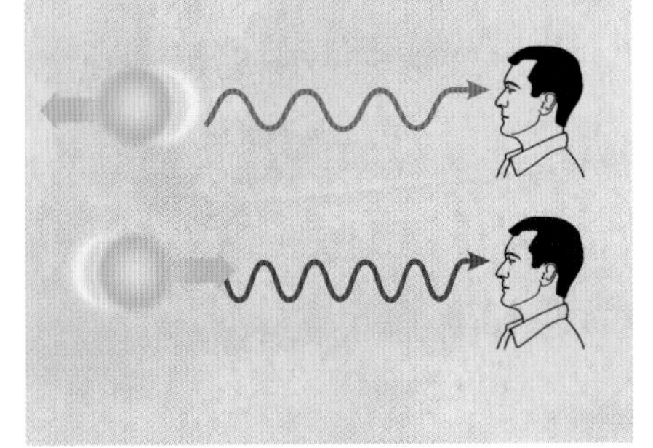

JUST AS POLICE SIRENS OR TRAIN BELLS CHANGE PITCH AS THEY WHIZZ PAST AN OBSERVER, a similar effect changes the color of stars, showing whether they are moving away or towards us. In 1929, it was found that they were all flying apart.

The phenomenon of the way a sound changes in pitch depending on the motion of its source relative to an observer is called the Doppler effect and was first described in the 1840s, taking the name of its Austrian proponent. Perhaps surprisingly from our modern standpoint where the audible manifestation of the effect is an everyday experience, Christian Doppler was an astronomer describing the light coming from binary stars in a rapid mutual orbit.

Andromeda approaching

In 1894, Vesto Slipher directed the large telescope at the Lowell Observatory in Arizona toward what was then known as the Andromeda Nebula. Slipher's boss, Percival Lowell, wanted to see if this and other spiral formations were in fact infant solar systems forming from a hot swirl of dust and gases. Slipher was using a spectrometer to look for tell-tale elements such as iron and silicon which might indicate a rocky planet forming. Instead Slipher found that the spectra had higher wavelengths than expected and correctly deduced that this was the Doppler effect of Andromeda hurtling towards us.

Everything moves

After two decades of surveys Slipher found that more distant galaxies, beyond our Local Group, were redshifted, meaning they were moving away from us, and from everything else—and fast; some speeds were 1,800 km/s (1,118 miles/s). In 1929 Edwin Hubble found that an object's redshift was proportional to distance. The Universe was expanding, getting larger as it got older.

Edwin Hubble surveyed redshifts in the late 1920s using the Hooker Telescope. Fitted with a 100-inch (254-cm) mirror and situated in clear skies 1,700 meters (5,577 feet) above sea level on top of California's Mount Wilson, this was the most powerful telescope of its day.

59 The Last Planet?

THE LOWELL OBSERVATORY WAS BACK IN THE SPOTLIGHT IN 1930 WHEN IT DISCOVERED WHAT LOOKED LIKE A NINTH PLANET. Perturbations in the orbit of Neptune suggested Planet X was out there.

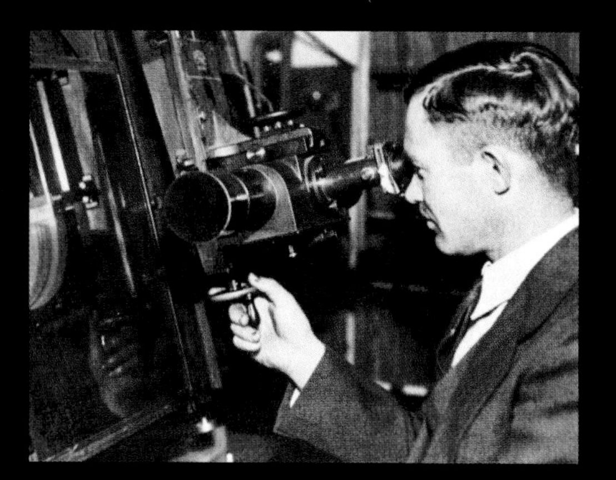

Clyde Tombaugh used a blink comparator to switch back and forth between photos—in the blink of an eye—and so show up any object moving position.

The Lowell Observatory began to survey the ecliptic for Planet X in 1906. In 1916, Percival Lowell died and his observatory began a dispute over funds with his widow Constance, which put a halt to the project until 1929. Clyde Tombaugh, a 23-year-old researcher, was then given the job of searching. He spent a year taking images of the sky every two weeks and comparing them to see if anything had moved. His discovery of a wandering object in 1930 made world headlines, and suggested names flooded in. (Constance Lowell proposed either Percival or Constance.) In the end the object was named Pluto, the god of the underworld, after an 11-year-old English schoolgirl suggested it was very cold that far from the Sun. However, after 76 years as the Solar System's ninth planet, tiny Pluto was demoted to the status of dwarf planet. We now know that Planet X does not exist.

60 The Death of Stars

BY THE 1930S, THE VERY HOT BUT VERY SMALL STARS, DUBBED WHITE DWARFS, WERE GETTING A LOT OF ATTENTION. They had been found to be very dense indeed, with their atoms packed closer together than was possible on Earth. The structure of this weird material got one Indian astronomer thinking while on a long ocean voyage to England.

Subrahmanyan Chandrasekhar would have known that the "degenerate matter" in white dwarfs was considered to be the end stage of a star, where immense gravity had crushed the remains into an ultradense star. The atoms inside were not held together by normal chemical bonds, but squashed together instead, only staying separate due to the repulsive force between their electrons. A white dwarf with the same mass as the Sun would be about the size of Earth, and stranger still, more massive stars would be smaller, not bigger.

Chandrasekhar calculated how massive a white dwarf could be before even the repulsive forces between electrons would not be enough to keep the electrons apart. The answer he got was 1.4 solar masses. He made this "Chandrasekhar limit" public in 1931 and sparked a debate about where stars bigger than this went.

Supernovas and neutron stars

One possible answer was that they collapsed into black holes, which, although predicted, were still wholly theoretical. However the math suggested that these were at least 10 times more massive than the Sun, so where did all the other giant stars go? In 1934, Fritz Zwicky and Walter Baade suggested that these giant stars died in a huge explosion, which they named supernovae. They then suggested that these events were the source of cosmic rays and the end product was a neutron star, comprised entirely of neutrons (atomic particles only discovered the year before). A neutron star with the mass of the Sun would be 12 km (7.45 miles) across! Baade and Zwicky used wide-angle telescopes to search for supernovae and found dozens, but the neutron stars stayed hidden for a few more decades.

Zwicky and Baade's theory was that the immense gravity inside a giant star collapsed the atoms inside into pure neutrons. That released a blast of energy like the one on the left, that could be seen as a new star, or a nova, only much more energetic. Hence it is known as a "supernova."

61 Dark Matter

MOST OF THE UNIVERSE IS MISSING. In 1932 Jan Oort found that the Milky Way was spinning too fast for the amount of material in it. Fritz Zwicky, seeing a similar effect in the motion of other galaxies, named the invisible material *dunkle materie*, better known today as dark matter.

Zwicky's suggestion was that the darkness of empty space was not that empty. Instead there was material that did not give out light and therefore could not be seen directly. Only its contribution to gravity could be detected. No one took much notice of dark matter—too hard to see it—for 40 years. Then in the 1970s, the amount of dark matter was measured by gravitational lensing light—measuring how it bends through space warped by matter. It was discovered that dark matter was five times more prevalent than plain old matter! No one knows what dark matter is. Two possibilities are WIMPs (Weakly Interacting Massive Particles)—they have mass but do not interact with detectors, or MACHOs (Massive Astrophysical Compact Halo Objects), which is a funny name for black holes, neutron stars, and brown dwarfs—stuff too dark to see. Of course it might just not be there at all.

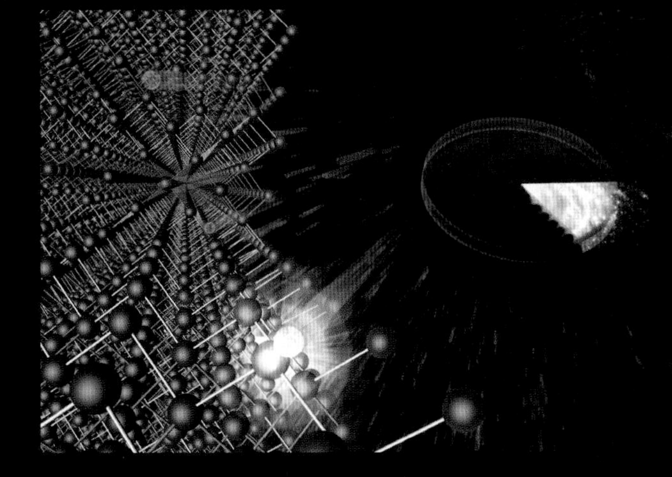

An illustrative attempt to show that which we cannot perceive: dark matter.

62 Solar Power

THE SUN HAS BEEN THE CONSTANT PRESENCE THROUGHOUT THE HISTORY OF EARTH, PROVIDING THE HEAT AND LIGHT THAT MADE LIFE POSSIBLE. Perhaps surprisingly for such a dominant feature, the workings of the Sun were something of a mystery until the development of quantum physics in the 1920s. Now we believe it to be a rather ordinary star.

It had long been known that sunlight appears to be white because it contains a full rainbow of colors—the spectrum, as Newton named it in the 1670s. In 1800, William Herschel repeated Newton's optical experiments, splitting sunlight into its constituent colors. He held a mercury thermometer—unavailable in the days of Newton—into the different colors to see how each contributed heat, and found that the thermometer

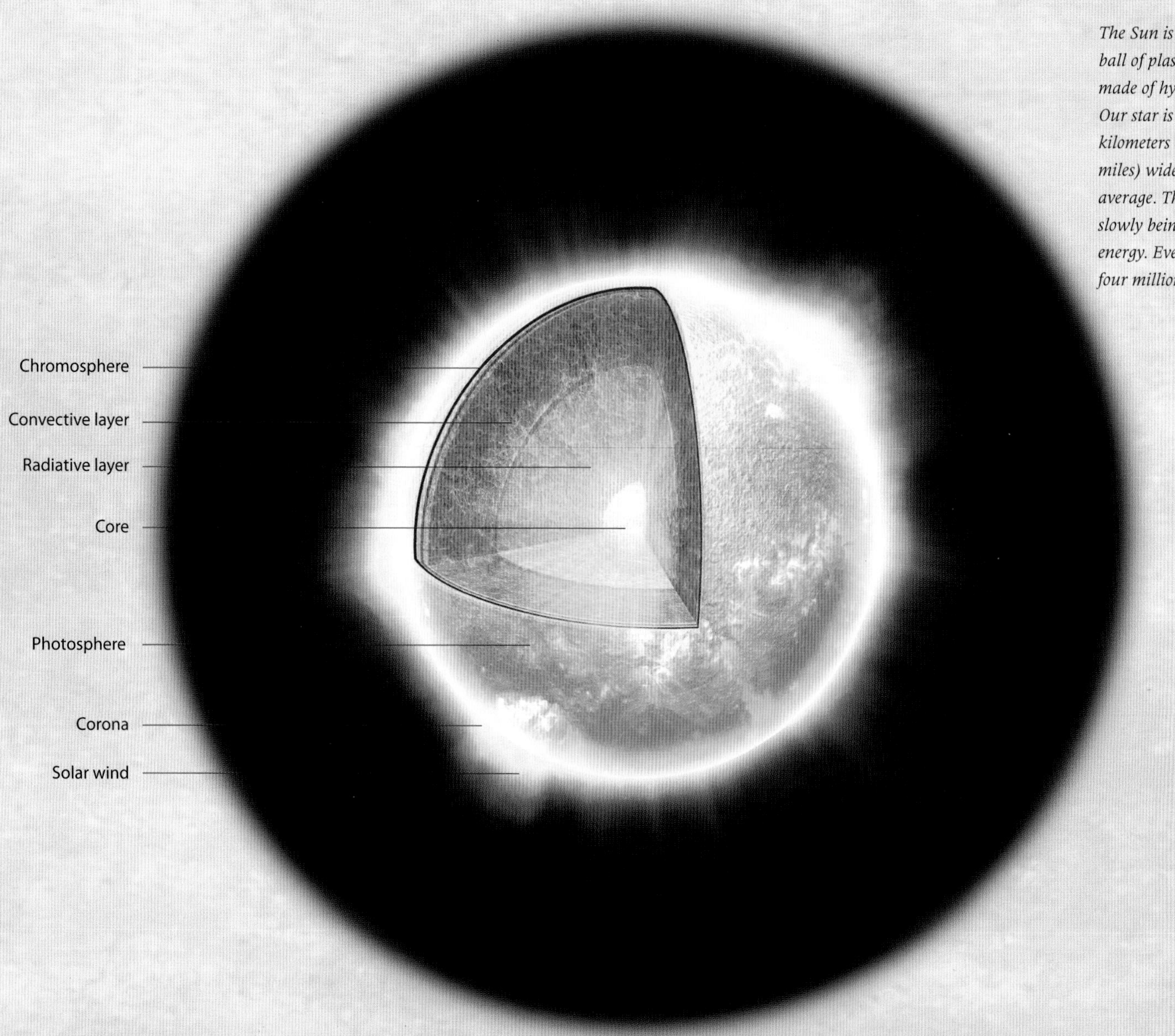

The Sun is an immense ball of plasma, mainly made of hydrogen atoms. Our star is 1.4 million kilometers (0.9 million miles) wide but that is very average. The Sun's mass is slowly being converted into energy. Every second it gets four million tons lighter.

increased in temperature fastest when held just next to the red—but not in it. The conclusion was that the heat of the Sun (and anything else) shines out as invisible infrared (meaning "below red") radiation.

From hot to cold

The laws of thermodynamics, which govern the ways energy behaves, say that heat energy always moves from hot places to cold ones, and so obviously the Sun is very hot. In the 1850s it was thought that the Sun was made from hot liquids. Lord Kelvin—a leading figure in thermodynamics—suggested that the source of the Sun's light was the effect of the fluids' huge gravitational energy being converted into radiation.

At the turn of the century, Ernest Rutherford, the godfather of nuclear physics, suggested that the heat was coming from radioactivity deep within the Sun. However, in the 1920s, Arthur Eddington, a towering figure in British astronomy, who had recently been lauded for helping Einstein prove the theory of relativity (his results were a bit off but history has forgiven him), entered the debate. He proposed that the atoms in the Sun would be under such force that the outer electrons would be ripped away, leaving a seething ball of plasma.

NUCLEAR FUSION

Hydrogen atoms have a simple structure—a negatively charged electron moving around a positively charged single proton. In the plasma of a star, the atoms are smashing into each other, so the protons and electrons separate. In normal conditions the protons' positive charges would make the particles repel each other, but in a star's core they collide with such force that they sometimes fuse. It is not as simple as two protons joining together to form the nucleus of a helium atom. Instead nuclear fusion goes through a number of steps, where a proton clubs together with neutrons (particles similar in size but without a charge) to form heavier forms of hydrogen. Two of these heavy forms, or isotopes, fuse into one helium nucleus (containing two protons and two neutrons). Where did the neutrons come from? When two hydrogen nuclei (each a single proton) fuse, one loses a tiny bit of mass and becomes a neutron. The lost mass is released as radiation and also as strange particles called neutrinos, which are very common but almost impossible to detect.

Squeezed middle

Helium had been found on the surface of the Sun, and was later shown to be a super-lightweight gas, heavier only than hydrogen. Eddington suggested that the helium was produced by hydrogen atoms fusing together, and this "nuclear fusion" was the source of the Sun's heat and light. At the time it was thought that metallic elements were the main constituents of stars, since they are what showed up clearly in spectrographs. In 1925, Cecilia Payne showed that hydrogen and then helium were present in far greater amounts in stars than on Earth. Finally in 1939 Hans Bethe, a German physicist, figured out the atomic steps by which fusion took place.

The great pressures required for fusion were only found in the Sun's core, from where energy radiates out, scattering around in all directions as it bounces off all the dense plasma. After thousands of years it reaches the outer, convective zone where it travels to the surface in vast upcurrents of hot plasma. Only then is the energy released as heat and light shining into space—arriving at Earth eight minutes later.

63 Space Bombs

PERHAPS UNSURPRISINGLY THE FIRST HUMAN-MADE OBJECTS TO LEAVE EARTH'S ATMOSPHERE were explosive missiles whose passage through space was an incidental part of their trajectories toward destroying their targets.

The pioneers of modern rocketry had their heads in the stars, but technology had its roots in weaponry, and as the cloud of war gathered once more throughout the 1930s the prospect of new weapons development arose once more. The Soviet rocket master, Sergei Korolev, was imprisoned by Stalin on trumped up charges, effectively pausing Russian missile technology. Robert Goddard had done much to develop liquid-fueled rockets, but the U.S. military did not make them a priority. Instead solid-fueled rockets were developed as less expensive and more effective artillery.

Decisive weapon

This was in contrast with the activities of Wernher von Braun, a young German rocket engineer who began his career as the assistant of Hermann Oberth, another leading figure in liquid-fueled rockets. Oberth was not von Braun's only mentor. Until the outbreak of war in 1939, the German—now a Nazi Party member like all leading academic figures—was in frequent contact with Goddard, picking his brains on steering and cooling systems.

The high costs that prevented other powers from developing large-scale rocket weapons were less of an issue for the German authorities, who had access to slave labor from their concentration camps. And as the tide of World War II turned against Hitler, he threw increasing resources behind decisive aerial weapons that could strike at the heartland of his enemies. The first was the jet-powered V-1 (or doodlebug), a pilotless, explosive-packed drone, that fell from the sky after a preset period of flight. However, the V-1 was vulnerable to anti-aircraft fire, so the V-2 rocket was introduced. This was a 14-m-tall (46 foot) vehicle that flew to an altitude of more than 100 km (62 miles) and had a range of 320 km (200 miles). It then fell to Earth at four times the speed of sound, far too fast to be detected and destroyed by air defenses of the day.

The first V-2 to enter space was launched in 1944 and soon after V-2s were attacking England, France, and Belgium. The rockets did indeed strike fear into Hitler's enemies but were not very effective weapons, each one killing an average of two people.

The V-2 was an extremely expensive weapon, costing more to develop than the atomic bomb of the Manhattan Project. After the war, submarine-launched V-2s were discovered, which were intended to attack the United States.

64 Rocket Man

THE BELL X-1 WAS A BULLET WITH WINGS, BUILT FOR ONE PURPOSE—TO FLY FASTER THAN SOUND. COULD A HUMAN SURVIVE SUCH A SPEED? It was going to take quite a pilot to find out. In 1947 Chuck Yeagar became that man. His survival opened the door to crewed rocket flights into space.

For his historic flight, Chuck Yeager renamed the X-1 "Glamorous Glennis" in honor of his wife.

The X-1 was a liquid-fueled rocket with wings and a cockpit. When it was built in 1945, jet-powered aircraft were few and far between. Rocket technology was way ahead. As far back as 1928 Alexander Lippisch, a German, had designed a glider body fitted with two solid-fueled rocket engines. A liquid-fueled rocket fighter, the Me-163 Komet ("Me" stands for Messerschmitt) saw action in the last days of World War II. It had a similar "flying wing" design to Lippisch's and could fly at 970 km/h (602mph), about the speed of a modern airliner. However it was hard to fly and stayed airborne for just a few minutes before running out of fuel—and it frequently exploded.

Pushing the envelope

The pilots of the Bell X-1 would have known the risks, too. The flattened wings of this aircraft made it cut through the air efficiently but did not make it fly well—if at all—at low speeds. As a result, the X-1 did not take off on its own but was dropped from a modified bomber aircraft (pilot and all) and the rocket engine was turned on as it fell.

The aircraft was tested by the Bell company on behalf of the U.S. military and NACA, (National Advisory Committee for Aeronautics, later to become NASA), with pilots gradually going faster and higher to see how the plane behaved. By October 1947, it was time to beat the sound barrier, and Captain Yeager was given the job. This was a flight into the unknown. No one even knew if the aircraft's control surfaces would work at this speed. Yeager lived to hear the sonic boom and tell the tale of being the fastest man alive. Later X-planes went faster and higher, to the very edge of space. Several of the pilots swapped flying suits for spacesuits and became the first astronauts.

65 Big Bang

IF THE UNIVERSE WAS EXPANDING WITH AGE, IT MUST HAVE BEEN SMALLER IN THE PAST. This suggests if we could run time backwards, all of space would shrink to a single point. Was that how the Universe began?

The Reverend Richard Bentley, a seventeenth-century clergyman in Worcester, England, was given the task of questioning Isaac Newton about his newfangled notion that the Universe was held together by gravity. Bentley did a very good job, asking why, if the Universe was finite and static, was gravity not pulling everything together into a final collapse. Newton's reply was that the Universe was infinite and therefore stable.

Dynamic not static

However, 250 years later, Albert Einstein was not content that he could configure a static Universe at all, finite or infinite. His successors, including another clergyman, the Belgian Abbé George Lemaître who also happened to be a leading physicist, realized that the only possible form of the Universe was a dynamic one—expanding or contracting. A contracting Universe seemed unlikely—it would have in all likelihood collapsed by now. Hubble's proof of an expanding universe in 1929 led Lemaître to propose that the dynamic Universe had begun in an almighty explosion. As much intuition as theory, Lemaître suggested a primeval atom shattered into all the atoms making up the observable Universe, which exploded outward from a single point in the distant past and had been heading that way ever since.

The idea gathered its supporters and detractors alike. One of the latter was responsible for giving it its current name: Fred Hoyle, an eminent astronomer, described it as a "big bang," while suggesting his alternative, the steady-state theory which proposed that matter is being added to the Universe continuously as it expands.

In 1948, a paper by Alpher, Bethe, and Gammow (a pun on the first three Greek letters) suggested that the Universe had developed through the continual fusion of primordial particles into more complex and massive forms. This process went hand in hand with the Universe cooling down and spreading out from a hot, dense past to a cool, diffuse future. Was there evidence for this tumultuous young Universe? Only time and better telescopes would tell.

The Big Bang is often characterized as an explosion, which leads people to imagine a bright light expanding outward from a point in blackness. The explosion actually happened everywhere all at the same time. It is just that everywhere—the whole of space—was a single point at the time.

66 Atom Factories

MORE THAN 99 PER CENT OF THE MASS OF THE SOLAR SYSTEM IS MADE UP OF THE SUN, AND THE GREAT MAJORITY OF THAT IS HYDROGEN AND HELIUM PLASMA. However, on Earth these elements are comparatively rare. Instead, elements with heavier, more complex atoms, such as oxygen, carbon, and iron dominate. Where did all these substances come from?

The Big Bang did not make atoms, not at first anyway. In the seething superheat of the infant universe mass and energy had yet to differentiate—it was all the same thing back then. As the Universe expanded, its contents spread out a little and cooled down, and began to form into subatomic particles, such as quarks and electrons—the stuff that atoms are made from—and an array of more exotic baubles, some charged, some not. Some were even "flavored" (in the jargon).

However, at the same time, the equal but opposite to this matter was forming too, as antimatter such as positrons (antielectrons) and antiquarks among other antiparticles. Matter and antimatter do not get on and annihilate each other upon meeting—resulting in radiation. Obviously the annihilation was not quite total. For reasons still not fully explained, there were more particles than antiparticles, so the leftovers became the Universe—the Sun, planets, you, me, and trillions upon trillions more solar systems. (It may be that there are regions of the universe made of antimatter but we have not seen them to date.)

PARTICLE ACCELERATOR

Scientific curiosity had resulted in people smashing atoms together to see if they fused several years prior to the first theories on their astronomical origins. Powerful electric fields were used to fire atomic nuclei at targets. The first accelerators were called cyclotrons. These devices sent small nuclei in a spiral path onto targets made from heavier atoms. This was how the first artificial elements were manufactured. Linear accelerators, like the one below from 1947, fired objects in straight lines. These devices were later developed for medical radiation therapy.

In the late 1940s, Fred Hoyle had discovered that heavier elements were more abundant in older galaxies than young ones, leading to the conclusion that not all atoms in the Universe were made all at once during the Big Bang.

Simple beginnings

So after this period of annihilation (it was all over in less than 10 seconds), the subatomic particles left behind began to live up to their billings and formed atoms. Trios of quarks formed protons. These became the first atomic nuclei, and after 370,000 years or so, the Universe had cooled down enough for the positively charged protons to bond with negatively charged electrons. Single protons each teamed with an electron to form the first atoms, all hydrogen. The energetic early universe fused some of these primeval atoms into helium, but the hydrogen made at that time is now spread across space. Even today three-quarters of all atoms are the hydrogen ones made in the Big Bang.

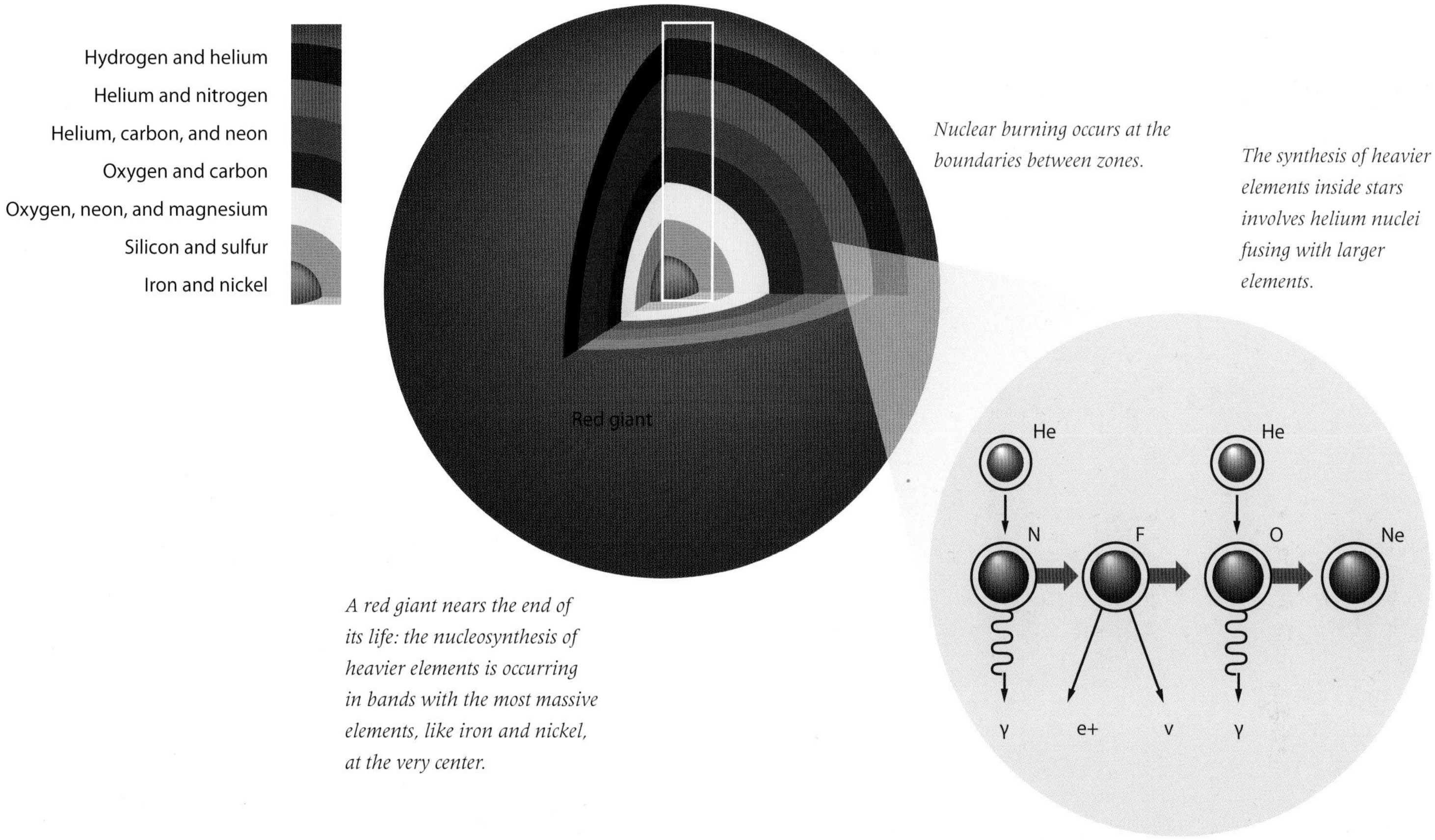

Nuclear burning occurs at the boundaries between zones.

The synthesis of heavier elements inside stars involves helium nuclei fusing with larger elements.

A red giant nears the end of its life: the nucleosynthesis of heavier elements is occurring in bands with the most massive elements, like iron and nickel, at the very center.

Inside stars

Over the next billion years, gravity pulled hydrogen into clouds, then spheres, which, once big enough to drive a fusing core, ignited into the first stars. In the 1950s, a team of four astrophysicists (star scientists) began to consider what happens inside a star over its lifetime. Geoffrey Burbidge, Margaret Burbidge, William Fowler, and Fred Hoyle, who have become known by their initials B^2FH, used computer simulations designed for testing nuclear weapons to investigate what was happening in stellar cores. In so doing they mapped out the fundamentals of nucleosynthesis, the process in which all atoms heavier than hydrogen and helium were made inside stars.

A star's supply of hydrogen fuel is not infinite and as it runs low, the star becomes predominantly helium plasma. Helium nuclei are roughly four times more massive than the original hydrogen, so they form a central fusing core. Any remaining hydrogen continues to fuse in shells around the helium. As a result the star's core grows larger and hotter, which in turn makes the rest of the star expand, forming a red giant. It is giant because it is hundreds of times larger than its original dwarf form, and it is red because its surface is cooler—the heat energy is spread more thinly over its huge surface. This is the fate of all stars, including the Sun in about five billion years.

In the red giant's core, three heliums fuse to make a carbon nucleus. When the helium runs low, the carbon is used to make oxygen, sodium, neon, which in turn create a whole family of heavier elements up to middleweight substances like iron and nickel. In most stars the fusion stops there, and the cooling star drifts away leaving only its hot core as a white dwarf. However supergiants will go supernova, and in the force of that explosion the heavy—and much less abundant—elements, such as gold, mercury, and uranium, are formed. As the singer said: "We are stardust."

67 Fellow Travelers

INTERNATIONAL GEOPHYSICAL YEAR IN 1957 WAS MEANT TO BE A WAY TO THAW COLD WAR TENSIONS, with collaborations between scientists from across the world, not least the United States and the Soviet Union. But the result was a renewed struggle, this time for control of outer space.

A replica of Sputnik 1—the real one burned up in the atmosphere after three months in orbit—shows what was inside the 58-cm (22-inch) sphere. The batteries lasted 22 days transmitting temperature data.

In the evening of October 4, 1957, a rocket took off from Site No. 1 at the Baikonur Cosmodrome in what was then deep in Soviet territory. Within a few minutes, the first artificial satellite was in orbit around Earth. The spacecraft was called Sputnik 1, which is simply the Russian word for "satellite," but might also be translated as "companion" or "fellow traveler." It weighed around 80 kg (176 lbs) and orbited every 90 minutes about 483 km (300 miles) above the surface—letting people know it was there with a tell-tale beep beep beep that could be picked up by amateur radio. Once sure Sputnik's orbit was stable, the Russian news agency TASS announced the event to the world.

Space race starting pistol

Western powers did not hear just a simple beep from space, they heard a declaration from the Soviet Union that their rocket missiles were the most powerful and reliable in the world, capable of launching science satellites into orbit—and perhaps also delivering nuclear weapons to any point on the globe. NACA lacked the means to track Sputnik and enlisted amateur astronomers to keep a lookout for it. Sergei Korolev, Sputnik's designer, had made the rocket booster, which also made orbit just in front of the satellite, highly reflective so it was easy to see at dawn and dusk.

The U.S. contribution to the International Geophysical Year was Explorer 1, launched in January, 1958 from a naval missile. Within months a new agency, National Aeronautics and Space Administration, or NASA, replaced NACA with the mission of catching up with the Soviets in the race into space.

THE SCIENCE OF A SPACE STAND OFF

Despite its scientific credentials, *Sputnik 1* did little more than measure the temperature of the spacecraft. *Explorer 1*, on the other hand, actually discovered something. Despite being a tiny 14 kg (31lbs), the spacecraft contained cosmic ray detectors. It was feared that the detectors were faulty since they reported large areas with no results and then small regions of dense activity. However, this matched the predictions of James Van Allen, who suggested that Earth's magnetic field would sweep cosmic rays into bands, which are most concentrated at the poles. The Van Allen belts funnel the solar wind toward the poles creating the light shows known as aurorae.

68 Space Animals

THE EXPLORATION OF SPACE GOES HAND IN HAND WITH A DESIRE TO SEND PEOPLE BEYOND EARTH. But before brave men made those momentous journeys, animals went before them.

At the dawn of the space race there were several unknowns that had the potential to prevent human spaceflight altogether. The escape velocity required for large enough craft is several times greater than the speed of sound. Could a body withstand the acceleration? Once in space radiation and extreme temperatures might kill the crew, while the heat produced by the friction of re-entering the atmosphere at great speed could literally cook them. And of course the rocket might blow up.

So the first Earthlings sent to space were monkeys called Albert, who were carried in Nazi V-2s captured by the U.S. military. None of them survived the test program in 1948 and '49. Albert I got to 63 km (39 miles) and suffocated—outer space starts at 100 km (62 miles). Albert II made it to 134 km (83 miles) and was doing fine until his parachute failed to open on landing (crashing). Neither Albert III or IV beat his record.

The first animals to make it home alive were two Russian dogs, Dezik and Tsygan in 1951, and another Soviet dog, Laika, was the first in orbit in November 1957 on Sputnik 2. There were no plans to bring her home and she survived six hours. Belka and Strelka on Sputnik 5 were luckier, returning unharmed from orbit in 1960. NASA tested their Mercury spacecraft in 1961 with a chimp called Ham. He was dressed in a spacesuit so was unharmed by a loss of air in the cabin. The stage was now set for humans to take the big leap and leave Earth behind.

Laika, meaning "barker" in Russian, was nicknamed as Muttnik by the less sympathetic Western commentators. She was fitted with sensors to show how her body responded to weightlessness. Although a poisonous meal was loaded on board to give Laika a painless death, she actually died from extreme heat in the cabin after a failure of the life-support system.

69 Space Diver

AS PILOTED FLYING MACHINES GRADUALLY EDGED CLOSER TO SPACE ITSELF, a new type of flight suit was developed—one that would later become the spacesuit. In 1960, one man went up to test it out.

An automatic camera captures Joe Kittinger as he begins the longest sky dive ever made. He was in freefall for 4 minutes 36 seconds and hit a top speed of 982 km/h (610 mph)—faster than a passenger jet. The chute system he used was a new design that stopped him going into a fatal spin.

Contrary to popular belief, the blood does not boil away when the body is exposed to a vacuum. In outer space you are more likely to freeze solid. In short exposures, a person loses consciousness in about 15 seconds and puffs up to twice their body size. In 1960, U.S. Air Force pilot Joe Kittinger traveled to a height of 31 km (19 miles) aboard a helium balloon. Although not quite space, the atmosphere up there is close to a vacuum, and Kittinger wore an early spacesuit with air kept at normal pressure inside. Nevertheless, a leak caused him to lose the use of his right hand. To get home Kittinger just jumped off. His parachute jump is still a world record.

70 Race to Space

WHILE OTHER NATIONS WERE ABLE TO LAUNCH SATELLITES, only the Cold War superpowers had ambitions of getting men into orbit. After the surprise head start of Sputnik 1, this race was going to be much closer.

By 1959 the Soviet space agency and NASA had assembled their candidates for human spaceflight. The Russian program was code-named Vostok, while the American initiative was Project Mercury. Both programs looked for people with a similar profile—not too tall or heavy for the capsule and capable of withstanding large G, low pressures, and whatever else the testers could dream up to whittle down the hundreds of applicants to a core team of super tough, super fit, little men.

Six men formed the initial Vostok group, becoming the first cosmonauts—the Russian term for astronaut. Project Mercury began with seven astronauts, who were all superb military pilots, with brains to match and perhaps a little wisdom being in their late 30s, ten years senior

The Mercury Seven look thoroughly futuristic in an old-fashioned way in 1960. Alan Shepard is back left, while Deke Slayton and John Glenn front and center have yet to be issued with proper space boots.

to their Vostok counterparts. The first cosmonauts were military men too but their prowess as pilots was less important. The Vostok spacecraft was highly automated, with the single crewman strapped into a reinforced sphere that had few means of steering from inside. The Mercury spacecraft, meanwhile, formed the cone at the top of the rocket, and had a window and steering controls so the astronaut within could actually fly his spacecraft.

In the end it was rocket technology that won the day for the Soviet Union, when Yuri Gagarin was launched into orbit aboard Vostok 1 on April 12, 1961. A month later Alan Shepard became the first Mercury astronaut, although the limitations of NASA rocketry could only send his Freedom 7 craft on a suborbital flight. In February 1962, the more powerful Atlas rocket put John Glenn into orbit (the first American to make this journey), but by now the Space Race had shifted to landing on the Moon.

In 2003, China became the third nation to put humans into orbit, with the launch of their "taikonaut" Yang Liwei. India plans to send a "gaganaut" to space in 2016.

71 Star Sailor

THE SPACE RACE WENT ROBOTIC IN 1962, WITH THE FIRST UNCREWED PROBES BEING SENT TO EXPLORE OTHER PLANETS. The first one was Mariner 2, which made a flying visit to Venus with some surprising results.

Contact was lost with Mariner 2 in January 1963. It remains in orbit around the Sun to this day.

The Space Age was all about thinking big, and the thinking was that one day—who knows, by the 1990s?—humans would be living on other planets. Although interplanetary exploration was very much driven by science, national prestige and the prize of extraterrestrial territory insured nations stayed in the Space Race. The Soviet Venera 1 probe was launched in 1961, but missed Venus completely. The following year a software bug sent NASA's Mariner 1 on a possible collision course with northern Europe so it was destroyed. Mariner 2 was more successful and reached Venus in December 1962. To reduce weight the craft had no retro rocket to slow it down, so the probe just hurtled past in half an hour. During the short visit the probe found that the Venusian atmosphere was more or less a constant temperature, suggesting that heat was trapped inside those thick clouds that made the planet sparkle so brightly in our sky. But how hot was it down there?

72 Eternal Echoes

IN 1964, TWO ASTRONOMERS USING A SUPERSENSITIVE RADIO ANTENNA TO TEST OUT THE LATEST COMMUNICATION SATELLITES stumbled across a signal that seemed to be coming from everywhere at once. The faint microwaves are now known as the Cosmic Microwave Background.

Wilson and Penzias inspect the metal funnel they used to detect the Cosmic Microwave Background. To reduce interference they cooled the electronics in their receiver with liquid helium to 4° above absolute zero.

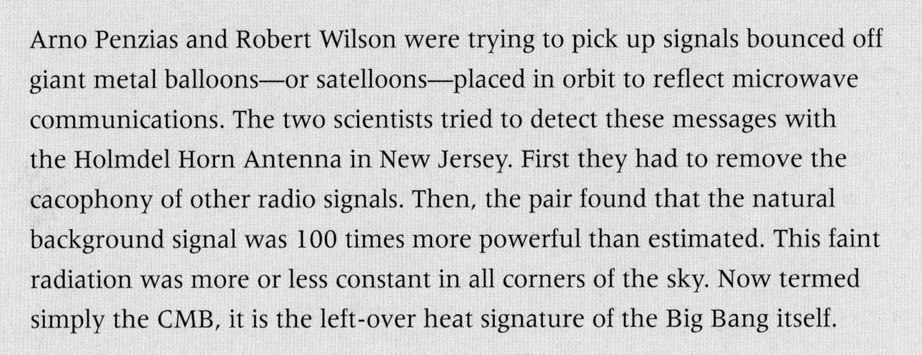

Arno Penzias and Robert Wilson were trying to pick up signals bounced off giant metal balloons—or satelloons—placed in orbit to reflect microwave communications. The two scientists tried to detect these messages with the Holmdel Horn Antenna in New Jersey. First they had to remove the cacophony of other radio signals. Then, the pair found that the natural background signal was 100 times more powerful than estimated. This faint radiation was more or less constant in all corners of the sky. Now termed simply the CMB, it is the left-over heat signature of the Big Bang itself.

73 Pulses from the Cosmos

IN 1967, WHAT LOOKED LIKE A FIELD FULL OF CLOTHES LINES picked up some regular radio pulses coming from space. No one really thought this was alien communications, but what else could it be?

Modern radio telescopes are made up of large antenna arrays which move together to always point at the same place in the sky.

Radio waves are radiation similar to light but with a lower energy and obviously invisible to the human eye. Astronomers had been exploring the radio waves coming from the sky since the 1930s. Radio telescopes are essentially large antennae, mostly dish-shaped for gathering faint signals. However, British astronomers Antony Hewish and Jocelyn Bell built a less dramatic looking radio detector in a field outside Cambridge. The Interplanetary Scintillation Array, as it was known, was designed to pick up fluctuations in signals.

Little green men?

In November 1967, Bell discovered a pulsating radio source that was eerily regular, always 1.3 seconds apart. The source moved across the sky with the stars, ruling out the possibility that it came from some secret artificial satellite or interference from earthbound radio stations. The most obvious explanation was that the radio pulses were produced by aliens, and Bell and Hewish labeled the source LGM-1 (Little Green Men 1). Then a second pulsing source was discovered in an unrelated corner of the sky, which put an end to the alien hypothesis.

The unusual bodies were termed *pulsars*—as in a pulsating star. It was suggested that the radio pulse was coming from one side of a rotating star, so it swept through the sky like the beam from a lighthouse, blinking on and off as it faced Earth. Pulsars spin very fast—some turn in a fraction of a second—and it was suggested they are all examples of neutron stars left behind by a supernova blast.

74 Gamma-Ray Burst

WHILE RADIO WAVES CARRY THE LEAST ENERGY IN THE RADIATION SPECTRUM, gamma rays contain the most. Just as faint radio pulses were revealing tiny stars across the sky, flashes of gamma rays were showing the most powerful events in the Universe.

It is estimated that a gamma-ray burst happens in the Milky Way every few hundred thousand years. It has been suggested that GRBs closer to home have caused mass extinctions on Earth.

Gamma rays are the kind of thing produced by nuclear explosions, and U.S. military satellites were stationed in orbit to look out for enemy nukes exploding in space in contravention of test ban treaties. In July 1967, two of the satellites detected bursts of unusual gamma rays, not at all in keeping with those produced by bombs, and from a source far beyond the Solar System. Cold-War logic led to this data being classified. But more advanced bomb-detector satellites found more and more of these gamma-ray bursts, typically lasting about 30 seconds.

In 1973 the data was made public but astronomers were largely stumped until 1991 when a space-based gamma-ray observatory was launched. This revealed that the faint sources were billions of light-years away. To be visible from that distance meant that a gamma-ray source released the same amount of energy in a few seconds that the Sun would give out in 10 billion years. The bursts may be neutron stars falling into black holes or the collapse of "hypergiant" stars hundreds of times the mass of the Sun.

75 Project Apollo

TWENTY DAYS AFTER LAUNCHING THE FIRST AMERICAN INTO SPACE, PRESIDENT JOHN F. KENNEDY PLEDGED TO PUT A MAN ON THE MOON BY THE END OF THE DECADE. The Apollo 11 mission fulfilled that promise with a few months to spare, giving the United States a monumental—but costly—win in the Space Race.

Neil Armstrong is all smiles inside Apollo 11's Eagle, the lunar module, after taking the first moon walk in history on July 20, 1969, watched by one-fifth of the world's population.

In today's money, the six Moon landings between 1969 and 1972 cost an eye-watering $18 billion each. But valuable lessons were learned about human space travel—not least to do it more cheaply next time—and technology spin-offs coupled with a Space-Age optimism kept the United States in the lead position in the Space Race until the end of the Cold War.

Project Apollo, named for the Greek god and archetype of masculine prowess, was grandly titled for its epic purpose. The missions to the Moon, both going around and landing on the surface, were the first and only times humans have left low-Earth orbit. The Apollo crews traveled nearly a million kilometers (more than 600,000 miles) at 32 times the speed of sound. Since the return of the final Apollo 17 mission in December 1972, no human has been more than a short distance above the surface of Earth.

Mission to the Moon

JFK announced Apollo just after the Mercury program had achieved its first crewed spaceflight. While Apollo began taking shape on the drawing board, five more Mercury flights tested NASA's ability to launch spacecraft and bring them and their occupants safely back to Earth.

This was followed by Project Gemini, with a new intake of astronauts, a spacecraft with room for two, and a more powerful rocket, Titan, to power heavier loads into orbit. The

Saturn V was the noisiest machine ever built and it remains the most powerful rocket to be launched successfully. The crew are sitting in the white cone, with only the narrow pointed escape rocket (for use in emergencies) above them.

The Apollo 11 lunar module heads down to its landing site at the Sea of Tranquility—chosen because it was flat. However, the lander was forced to cross a large crater before touching down and almost ran out of fuel.

Gemini missions were designed to test how long astronauts could function in space. One crew stayed in orbit for just shy of two weeks. On another trip, astronauts experimented with spacewalks, or extravehicular activity. The EVAs helped to show the limitations of operating in a full spacesuit, and to test tools that might be required for making repairs. Finally, Gemini pilots, including one named Neil Armstrong, practiced steering the spacecraft in orbit and docking it with another. They used an unmanned target vehicle called Agena for this, but the astronauts would have known by this time that when it came to Apollo these maneuvers would be crucial for getting to the Moon.

Preparing the way

While the crew vehicles were being designed, NASA sent probes to the Moon to see how the land lay. The first were Rangers, literally fired into the Moon in 1964, sending back pictures before they crashed. Next, five orbiters were sent to search for likely landing zones, and finally seven Surveyor robotic landers touched down between 1966 and 1968.

Once again the Soviets had got there first; Luna 2 had impacted the Moon in 1959, while Luna 9 made the first soft landing on a heavenly body a few months prior to the arrival of Surveyor 1. But from then on, all Soviet Moon probes, which continued into the 1970s, were eclipsed by the events of the Apollo program

The Soviet Union failed to develop a rocket powerful enough to lift a manned spacecraft to the Moon. But for Apollo 11, NASA's Saturn V rocket carried a crew of three in a service module (SM), which traveled to the Moon over three days, pushing a lunar module (LM) before it. Two crew members transferred to the LM and flew to the surface, before rendezvousing in lunar orbit with the service module, which carried them homeward. Finally, the crew cabin detached from the SM, forming a heat-shielded command module, which was all that made it back to Earth. Just 12 men have walked on the Moon. The last was Eugene Cernan, his parting words: "We leave as we came and, God willing, as we shall return, with peace and hope for all mankind."

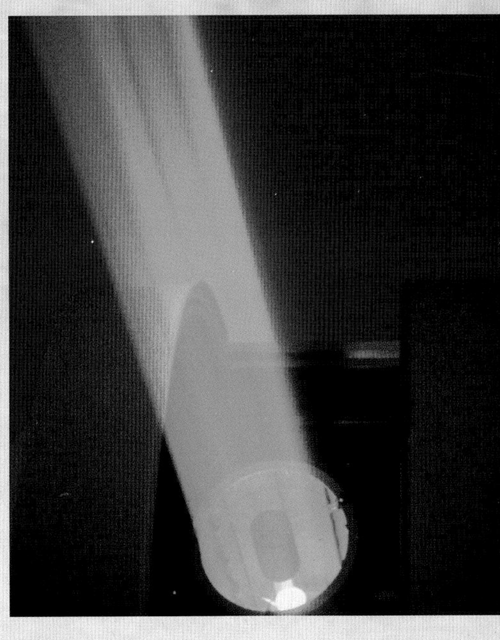

REAL OR FAKED?

A popular conspiracy theory has it that the Moon landings were filmed in a studio. However, the Apollo astronauts left mirrors on the Moon's surface, which reflect lasers fired at them from observatories in France and Texas. The reflections are used to provide accurate distances to the Moon. In 2009, a new orbiter surveyed the Moon and sent back pictures of the Apollo 17 landing sight, complete with American flag.

THE NEXT FRONTIER

76 Space Stations

WHILE NASA WAS STILL AIMING FOR THE MOON, THE SOVIET SPACE AGENCY HAD CHANGED ITS FOCUS. If humanity was going to explore space then we needed to learn how to live in it. In 1971, the first space station was launched, orbiting a collection of laboratories where the main experimental subjects were the crew members themselves.

The human body is built to resist gravity, and without it, the bones and muscles begin to thin. The crew of space stations must exercise regularly to combat this atrophy. In space the heart also pumps larger amounts of blood to the head—gravity pulls it down again on Earth—but in orbit it makes the face swell slightly and may damage the eyes.

The early space missions, not least the Moon shots, had perfected the technology of maneouvering craft in space to precise positions and systems had been developed for moving crews from one craft to another. The next thing to test was the human body itself to see how it fared in long periods of weightlessness aboard what had already been dubbed in science fiction as space stations.

The first space station was Salyut 1, launched empty in 1971. A crew of three followed in a Soyuz spacecraft but was unable to dock with the station and so returned to Earth. A second crew was more successful and spent 23 days in orbit, a world record at the time. However, tragedy struck on their reentry when the Soyuz craft crashed killing all crew inside. Salyut 1 was abandoned but what had been learned about its life support systems was applied to a succession of several more Salyuts over the next

decade. In that time NASA launched only one space station, Skylab, built inside the final stage of a Saturn V rocket. Despite some success, the American crewed space program became focused on the shuttle.

Living in space

In 1986 the Salyut 8 mission was transformed into Mir ("peace" in Russian), which was the first modular space station. Over ten years, a total of five modules were connected to the central orbiter, and two docking ports for crew vehicles and supply craft made it possible for the station to be occupied more or less permanently.

Valeri Polyakov looks out of Mir in 1995. This cosmonaut holds the record for the longest period in space, spending 437 consecutive days in weightlessness.

Despite suffering a fire, collisions, and meteorite strikes, Mir stayed in orbit until 2001, welcoming cosmonauts from around the world. Its crews stayed in space for months, if not years, showing how the body (and mind) coped with weightlessness—essential information if humans are ever to travel beyond the Moon.

77 Sagittarius A*

WHAT WOULD HAPPEN IF ONE BLACK HOLE SUCKED IN ENTIRE STARS AND EVEN OTHER BLACK HOLES? The result would be a supermassive black hole, which given enough time would become millions of times heavier than the Sun. In 1974, astronomers found one at the center of the Milky Way.

The first evidence of the huge black hole was a radio source coming from Sagittarius, the constellation containing Galactic Central Point—around which the Milky Way rotates. The pulse was pinpointed as originating in a dense region of Sagittarius A—a very active patch of sky—and was named Sagittarius A*. Conclusive evidence that this object is a supermassive black hole remained elusive since the radiation coming from that area was easily lost as it passed through many bands of stars and dust in between. Other large galaxies appear to have supermassive black holes at their cores, and many think that all middle-aged galaxies (like the Milky Way and Andromeda) are held together by big black holes. In 2008, after a 16-year survey, Sagittarius A* was confirmed as being 4 million times the mass of our Sun.

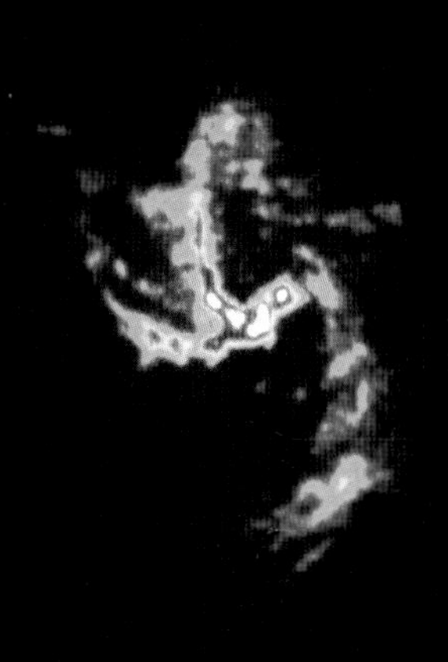

This image of Sagittarius A is formed from radio waves produced by ionized gases in the central region of the galaxy. The black hole A is near the middle.*

78 Touching Down

PLANS TO SEND HUMANS TO OTHER PLANETS HAVE NEVER BEEN ANYTHING MORE THAN THAT, JUST PLANS. A JOURNEY OF THAT DISTANCE IS A DAUNTING PROSPECT IN TERMS OF COST AND TECHNOLOGY, but that has not stopped us sending landers to take a ground-level view of these alien worlds. What we saw was frequently a surprise.

The Soviet space agency led the way in interplanetary exploration, not always successfully. The first craft to land on another planet, admittedly rather heavily, was Venera 3, which crashed into the surface of Venus in 1966—all according to plan. However, as soon as the probe vanished into the thick Venusian clouds, its detectors failed so space scientists were none the wiser from the visit. The following year Venera 4 was parachuted down to the surface, sending back readings of air pressures dozens of times higher than Earth and oven-scale temperatures. The probe was crushed into junk before it reached the ground. In 1970 Venera 7 was built to survive the conditions. It landed with a bump

Weighing in at five tons, Venera 9 proved tough enough for Venus, just. The mission almost failed when one plastic lens cap melted over the second camera.

Viking 1 sent data from its Martian landing site at Chryse Planitia (Plains of Gold) for more than six years. Researchers became quite attached to the view the probe presented, naming the large rock in the foreground Big Joe.

and promptly fell over, but managed to send back data for 23 minutes. Five years later, the Soviets thought they had left nothing to chance with Venera 9. This was the first craft to send back photos of another planet. For 53 minutes before succumbing to local conditions, the lander revealed a barren, desert world with a crushing atmosphere that would squash a human body and heat that would roast the mangled remains. Understandably, attention had already turned to Mars.

The Vikings land

NASA scored the first wins with the red planet. In 1971, their Mariner 9 arrived at Mars and became the first probe to orbit another planet. The survey it took revealed some huge landscapes, such as the Tharsis Bulge and the Valles Marineris, a "grand canyon" wider than the entire lower 48 U.S. states.

In 1976, NASA also put two Viking probes into orbit around Mars. These then dropped landers which plummeted into the Martian atmosphere. This is considerably thinner than Earth's, but frictional heating required the Viking craft to be protected by heat shields. Viking 1 parachuted the final part of the journey, touching down a couple of months before Viking 2. Both probes worked perfectly, sending back not only data from onboard instruments that analyzed the soil but also the first pictures, greeted with fanfares and with great excitement. There was initial confusion about coloring but the Vikings revealed that Mars's pale blue sky was frequently tinged with pink from the iron-rich dust that covers the red planet.

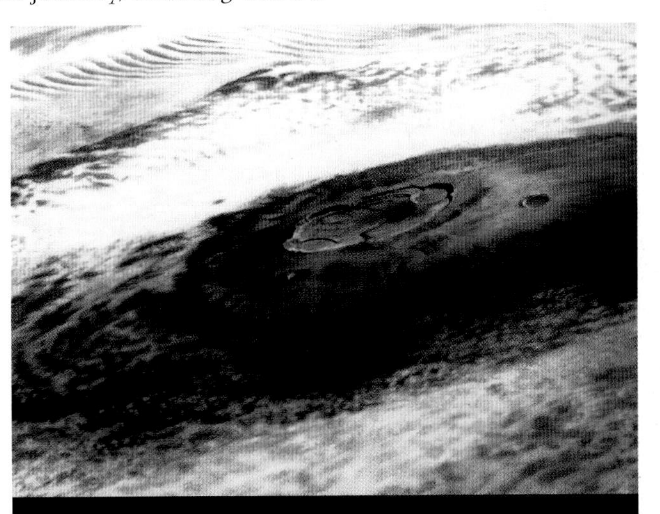

SMALL PLANET, BIG LANDSCAPES

The Mariner 9 survey and others since revealed that Mars is a volcanic planet but it does not have the tectonic plates of Earth. Earth's plates are constantly shifting around, buckling against each other and making even the largest mountain range or ocean basin a transient surface feature. However, once a feature forms on Mars it keeps on developing for millions, if not billions, of years. The Tharsis Bulge runs along Mars's equator and covers about a quarter of the planet. Astronomers suggest it is formed by magma pushing up from inside. This force creates rifts, such the Valles Marineris, and the magma feeds several huge volcanoes. The most spectacular one is Olympus Mons (above), which is 27 km (17 miles) above the surface, built up by innumerable eruptions which have made it the tallest mountain in the Solar System. The volcanoes on Mars have been dormant for 150 million years but will erupt again.

79 Studying Moon Rock

THE MOON IS THE ONLY HEAVENLY BODY WITH SURFACE FEATURES VISIBLE FROM EARTH WITH THE NAKED EYE. The modern study of the Moon has its very own science—selenology. Space travel means that selenologists can get "hands on" with the Moon—its rocks anyway. The last batch of Moon rock was brought to Earth in 1976.

The Moon is a constant companion, always showing the same face to us as it progresses through the sky night or day. However, one should not think that the Moon is static. It does rotate around its own axis, but in the distant past that rotation has become locked with that of Earth. The time it takes for the Moon to move around Earth is equal to the time it takes for the Moon to rotate on its axis once—and so although both bodies are in a constant spin, the same side of the Moon is always locked toward us.

This effect is called tidal locking and is the result of the gravity of Earth causing the entire Moon to bulge toward us ever so slightly. This is precisely the same effect that the Moon's gravity creates in the oceans, forming a bulging tide. Just as that bulge sweeps around Earth each day, so did the rock bulge on the Moon. That had the effect of gradually slowing the rotation of the Moon, until the bulge maintained a constant position relative to Earth, locking the Moon in place.

Lunar missions have returned with almost 400 kg (880 lbs) of Moon rock. It is a type of basalt, a rock of volcanic origin which forms when molten lava cools and solidifies quickly.

Of course our image of the Moon does change, and its position relative to the Sun is not locked. We can only see the Moon when it is lit by sunlight, and so our view of it waxes and wanes between the disc of a full moon to the crescent of a new moon as it is illuminated from different angles.

Lunar seas

The most obvious features on the lunar surface are dark regions, which early observers assumed were bodies of water, and named maria (singular, mare), the Latin for "seas." The lunar maria were then given fanciful names, such as Ocean of Storms, Sea of Serenity, and Bay of Rainbows. The maria are not water. If that does exist on the Moon's surface it will be as highly rare patches of ice lurking in deep craters. The maria are flat plains formed by ancient volcanic eruptions that flooded lowland regions. Despite dominating our view of the Moon, maria only cover about 16 per cent of its surface. It is not clear why this is; it may just be that magma was pooled on the near side. Whatever the reason, it is all over now. The maria are at least a billion years old.

Ups and downs

Before Galileo turned his telescope to the Moon in 1609, it was assumed that it was a smooth orb. However, Galileo saw a rugged world, with mountain ranges and craters. The pale regions of the Moon became known as the highlands and were named after

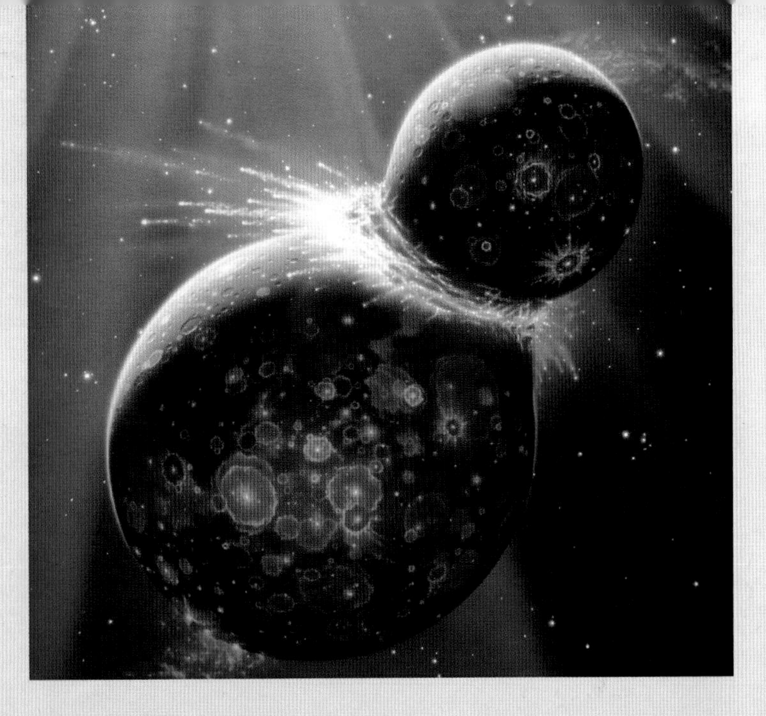

An artist's impression shows Theia—the theoretical planet that hit the young Earth to create the Moon. Theia is the mother of the Moon goddess.

Earth's mountain ranges. (Early selenologists found it easier to see surface features when they were cast in the shadow of the terminator—the line between day and night running across the Moon.)

In the 1650s, Giovanni Battista Riccioli named a large lunar crater for Copernicus, and the tradition of honoring great astronomers in this way has remained. Galileo suggested that craters were volcanic in origin, but closer studies showed overlaps and debris fields that could only result from meteorite impacts. With no air or water, the lunar surface is not eroded and so the craters—most more than 3 billion years old—have not aged a bit. (The lunar soil, or regolith, is the dusty remains of Moon rock pounded by countless impacts.)

Where did the Moon come from?

After the first few pieces of Moon rock had splashed down with Apollo 11 in 1969, geologists found they were remarkably similar to those in Earth's crust. The only real difference was a dearth of heavier metals, present in Earth's deeper regions. This suggests that the Earth and Moon formed from the same thing. One possibility is that a Mars-sized planet hit Earth more than 4 billion years ago. The collision melted Earth's crust and flung a good deal of molten material into orbit—where it coalesced into the Moon.

The near and far sides of the Moon look very different. The near side is less marked by impacts, protected as it is from bombardments. The far side is frequently called the dark side. However the Sun shines as much here as on the near side. We just can't see it.

80 Voyager 1 and 2

IN THE SUMMER OF 1964, A YOUNG ENGINEER WORKING FOR NASA BETWEEN COLLEGE DEGREES was given the job of investigating launch windows for missions to the giant planets. Gary Flandro saw in the orbital data the chance of a lifetime—the opportunity to send probes on a grand tour of the Solar System, taking in all four outer planets.

Three moons of Jupiter, Europa, Calisto, and Ganymede, may have liquid water and could support simple aquatic life, such as exists in the deep sea on Earth. In 2030 a European mission called Juice will return to the moons to search for these aliens.

If a Voyager craft is found by an alien race in the future, a gold-plated video disc on board carries photos of the Earth, a greeting from Jimmy Carter (then President of the United States) among many others, and a sound medley including whale song, Mozart, and Chuck Berry's "Johnny B. Goode."

The launch window was set for the late summer of 1977, and the tour would continue until 1989. Over this extended period, probes would harness the gravity of the giant planets along the way to fling them at their next target. All this was possible because of a rare alignment of the outer planets that meant they were all in the same part of the Solar System for a few years.

Given the go-ahead in 1972, the program was named Voyager. Two probes were planned, only one of which would make it all the way to Neptune. A couple of Pioneer probes—NASA's off-the-shelf planetary explorers—were launched to check out the route. Pioneer 10 swung by just Jupiter, while Pioneer 11 became only the second spacecraft to use a "gravity assist" from Jupiter to fling itself towards Saturn.

The Voyager probes were three times the size of the Pioneers and were equipped with cameras, spectrometers, and cosmic ray detectors. The caprices of interplanetary travel meant that Voyager 2 took off several weeks before Voyager 1, but by September 1977 the stage was set.

Among dozens of discoveries, Voyager 1's cameras captured volcanic eruptions on Jupiter's moon Io. The gravity of the giant planet churns up the inside of the moon, making it melt. Eruptions throw plumes of lava 150 km (93 miles) above the surface, making Io the most volcanic place in the Solar System.

Visiting the giants

Voyager 1 traveled faster than its sister probe and arrived at Jupiter first in January 1979, getting a close look at a swirling atmosphere and discovering a faint ring system around the gas giant, before being flung off to Saturn, arriving 18 months later. Here Voyager 1 was sent to look at Titan, the largest moon and the only satellite body in the Solar System with a thick atmosphere.

The Titan trip meant Voyager 1 ended its grand tour there and headed out into interstellar space. Meanwhile Voyager 2 had flown by Europa, Jupiter's ice moon, and was able to continue past Saturn in 1981, Uranus in 1986, and then Neptune in 1989. The probe imaged many of their moons for the first time and found the last planet also had rings, before it too began a journey into deep space which continues to this day.

The Voyagers are the most distant spacecraft from Earth but are still operating; the nuclear power sources will last until at least 2025. Astronomers are waiting for them to cross the heliopause in the next few years, the point where the solar wind, the charged particles streaming out of the Sun, become undetectable. The last image from Voyager 1 was received in 1990 showing Earth as an incredibly faint, pale blue dot against the empty black of space.

81 Magnetic Stars

SMALLER THAN MANY OF EARTH'S CITIES, WEIGHING MORE THAN THE SUN, AND WITH A MAGNETIC FIELD so strong that any object within 1,000 km (621 miles) is ripped apart. This is a magnetar—an exotic type of neutron star, first detected in 1979 but still barely understood.

The first neutron stars detected were radio pulsars, sending out beams of radio waves as they spin around at incredible speeds. Was it possible that pulsars sent out other forms of radiation? In 1979, the Venera orbiters, used to deliver probes to Venus, detected a short gamma-ray burst 2,000 times higher than the norm. Over the next few seconds, the radiation rippled through spacecraft at work on different jobs in Earth's neighborhood. What was being detected was the shockwave of a supernova from 5,000 years before. Supernovae produce neutron stars, made from the sci-fi friendly substance neutronium, matter so dense it has degenerated into pure neutrons—a teaspoon of this stuff weighs 200 million tons. Then astronomers found a pulsar that released high-energy gamma rays instead of radio pulses. Its magnetic field is 300 million times stronger than Earth's, possibly produced by the combination of its high temperature and relatively slow spin speed.

82 Shuttle: Reusable Space Plane

BY THE 1980S, SPACE TRAVEL WAS AS MUCH ABOUT BUSINESS AND SCIENCE AS IT WAS ABOUT DEFENSE—BUT STILL WITH A LARGE DOSE OF NATIONAL PRESTIGE THROWN IN. NASA MET THIS NEW ERA WITH A NEW SPACECRAFT, the reusable space shuttle, which blasted into space like a rocket and then flew home like an airplane and was still the heaviest air transport around.

Few of the early shuttle design concepts took account of the conflicting requirements of aerodynamics, weight restrictions, and payload capacity. The resulting orbiter was described as a "flying brick."

As the great expense of Project Apollo began to mount up, NASA was instructed to find a way of making space travel a lot less expensive. The solution was the rather prosaic sounding Space Transport System (STS). It became known the world over as the space shuttle. The space shuttle defined a new era for NASA, a new high profile success after the triumphs of Apollo. It also redefined the space program and would dominate it for the next 30 years.

Space truck

The plans for the space shuttle were seeded in the months before the Apollo 11 moon landing. A prescient thinktank within NASA suggested that a future spacecraft should not only be cheaper to launch but also be flexible enough for use by space scientists and the military, as well as being hired out to business as a space truck hauling commercial satellites into orbit.

The now familiar white space plane, or orbiter in space-speak, was the central component in this but one that could not get into orbit on its own. The multistage concept, used in earlier heavy lifting rockets, was modified for the STS. The orbiter carried little of its liquid hydrogen and oxygen fuel itself and was supplied during the launch from an external

SPACESHIPONE

While cheaper than Apollo's $18-billion-per-flight price tag, it still cost NASA $450 million to launch a shuttle and so yet again less expensive spacecraft were investigated. In 2004, the first privately owned spacecraft was launched—and it went to space twice in a fortnight, thus winning the $10 million X Prize created eight years before. SpaceShipOne (left), as the craft was called, powered to an altitude of 100 km/62 miles (where space begins) with a rocket engine. However, it launched from beneath a high-flying jet-powered mothership. This mode of spaceflight is now being developed to take tourists into space.

tank. The orbiter rode the large tank to the edge of space before jettisoning it. The tank burned up on reentry, the only component not to be reused in some way. The final pieces of the launch jigsaw were the solid-rocket boosters, essentially the largest fireworks ever made, which gave the shuttle that extra boost to get it skyward. These were dropped from the assembly two minutes into the flight and parachuted back down the 45 km (28 miles) for refueling and reuse.

The next step

The first shuttle was named Enterprise but was built for atmospheric testing in the late 1970s. The first space-ready version was Columbia, which launched in 1981. Four more followed, Challenger, Discovery, Atlantis, and Endeavour (all named for famous ships). The later orbiters had lighter-weight designs and so could carry slightly more cargo into orbit.

The shuttles could launch satellites into low- or high-Earth orbit. They could retrieve defunct satellites, launch space probes, carry laboratories for experiments in weightlessness, and pay visits to space stations. They were so useful the Soviets made a copy called Buran, which flew only once in 1988. Despite hundreds of successes, space flight never quite became routine and two shuttles were destroyed on missions. The last shuttle flew in 2011. A military drone, called the X-37, is the only reusable space plane currently in service, and the world waits for the next step in space travel.

The main body of the shuttle orbiter was the payload bay capable of carrying up to 30 tons of satellites and science equipment into low-Earth orbit.

83 The Great Attractor

EARTH GOES AROUND THE SUN, AND THE SUN ORBITS THE CENTER OF THE MILKY WAY. The galaxy is on the move too, gradually clustering together with other members of the Local Group, but that cluster also appears to be being pulled on by a dark, mysterious—and huge—mass.

Hubble's Law on the expansion of the Universe says that distant objects have a redshift because they are moving away from us, the observers. The space their light is moving through is expanding, stretching its wavelengths. All distant galaxies are also redshifted to each other—everything is moving away from everything else. However in 1986, a survey of redshifts found that their expansion was not constant in all directions. A gravity anomaly akin to a mass of 10,000 Milky Ways was evidence that expansion was not uniform. The Great Attractor, as this anomaly was named, is still a mystery, but many of its effects have since been attributed to a massive cluster of galaxies in space way beyond its location. The Universe appears rather uneven.

84 Comet Encounter

THE YEAR 1986 SAW THE FIRST VISIT OF HALLEY'S COMET SINCE THE START OF THE SPACE AGE—and a fleet of probes was sent to take a look at our familiar visitor.

Famously the orbital period of Halley's Comet had been calculated by Edmond Halley (who else?) centuries before so it was no surprise that the comet was coming. As it happened, this time around the comet passed Earth at a greater than usual distance and so few Earthlings witnessed the famous tailed star firsthand. But the world's space agencies were ready to get a closer view than ever before.

A fleet of five probes was launched to meet up with the comet in the middle of March that

Giotto was a modified design of a research satellite. It was fitted with a dust shield to protect it from the debris around the comet. Much of the shield was made from kevlar, the material in bulletproof vests.

THE ORIGINAL GIOTTO

Halley's Comet had made frequent appearances in culture though the ages. It is by far the brightest short-period comet. That means it comes around regularly enough to be recorded against the average span of human life, while long-period comets may only appear once every few centuries, millennia, or even longer. One of Halley's Comet's most considered appearances is as the Star of Bethlehem in the Adoration of the Magi, a painting of the birth of Christ made in 1304 by the Italian artist Giotto di Bondone—for whom the 1986 probe was named. Giotto had seen Halley's Comet during its visit in 1301 and what he saw, a bright ball with long tail, is apparent in his depiction of the astronomical object that led the Magi (three kings or wise men), from Persia to Judea.

year. Rivalries were put aside and mission profiles were synchronized to insure the best results. NASA had intended to observe the comet from the shuttle but that was canceled following the loss of Challenger six weeks before. However, NASA repurposed an earlier probe, International Cometary Explorer, to take a large-scale view of the comet, which by now had a tail longer than the width of the Sun. Two Japanese probes, Suisei and Sakigake, also kept their distance, studying how the passage of the comet affected the space around it. Two Soviet Vega probes arrived from dropping off landers at Venus to pass within a few thousand kilometers/miles of the comet's solid nucleus, and photographed its coma, or envelope of glowing gas. The best was left till last when the European probe Giotto plunged straight into the coma and passed within 596 km (370 miles) of the nucleus. Just 14 seconds away from its closest position the probe lost contact, but only after sending back some valuable data. (Mission control eventually turned it back on and Giotto visited another comet in 1992.)

Giotto got close enough to take photographs of the nucleus of Halley's Comet itself.

What the probe saw

Before going off-line Giotto confirmed predictions that the comet was a huge "dirty snowball," a lump of ice and rock about 16 kilometers (10 miles) long and 8 kilometers (5 miles) wide, all coated in a thick layer of ultrafine dust. When the comet is near the Sun, the solar wind, a high-energy stream of particles released by the star, flows around the nucleus, causing it to heat up. Gases erupt from fissures on the comet's surface and form a glowing trail of gas, plasma, and dust pushed along by the solar wind. Somewhat counterintuitively, a comet's tail always points away from the Sun, even when it is hurtling away from it. Once away from the Sun, the comet goes dark once more.

Giotto had been knocked out of action by a tiny grain of dust blasted from the comet. Analysis of the comet's dust showed that the nucleus was 4.5 billion years old, a left-over from the formation of the Solar System.

85 SN 1987A

THIS INNOCUOUS CODE REFERS TO A MOMENTOUS EVENT IN ASTRONOMY, THE FIRST SUPERNOVA TO BE OBSERVED IN ACTION. On February 23, 1987, the light from a giant exploding star reached Earth. The star had died long before the dawn of civilization, but in a twist of fate its emissions only arrived after astronomers were able to understand what they were seeing.

When a large star reaches the end of its life it goes out with a bang. The bright light it releases forms a new light in the sky, generally termed a *nova*. However, not all novae are dying stars, so astrophysicists clarify these most giant of explosions as *supernovae*. It is estimated that a big star goes supernova every 50 years in our galaxy. One was recorded in 1604—but then nothing for centuries. Then at 7.53 GMT, February 23, 1987, 24 antineutrinos—infinitesimal particles produced when atoms are being ripped apart—were detected in labs around the world, a huge spike in the normal level. It was calculated that these were just a tiny handful of the staggering 10^{58} neutrinos thrown out in all directions by an explosion. Three hours later the first starlight of this event arrived, 168,000 years after it was released. It was visible to the naked eye for a few months, and views through the best telescopes of the time revealed a glowing ring of plasma illuminated by the flash of the exploding star.

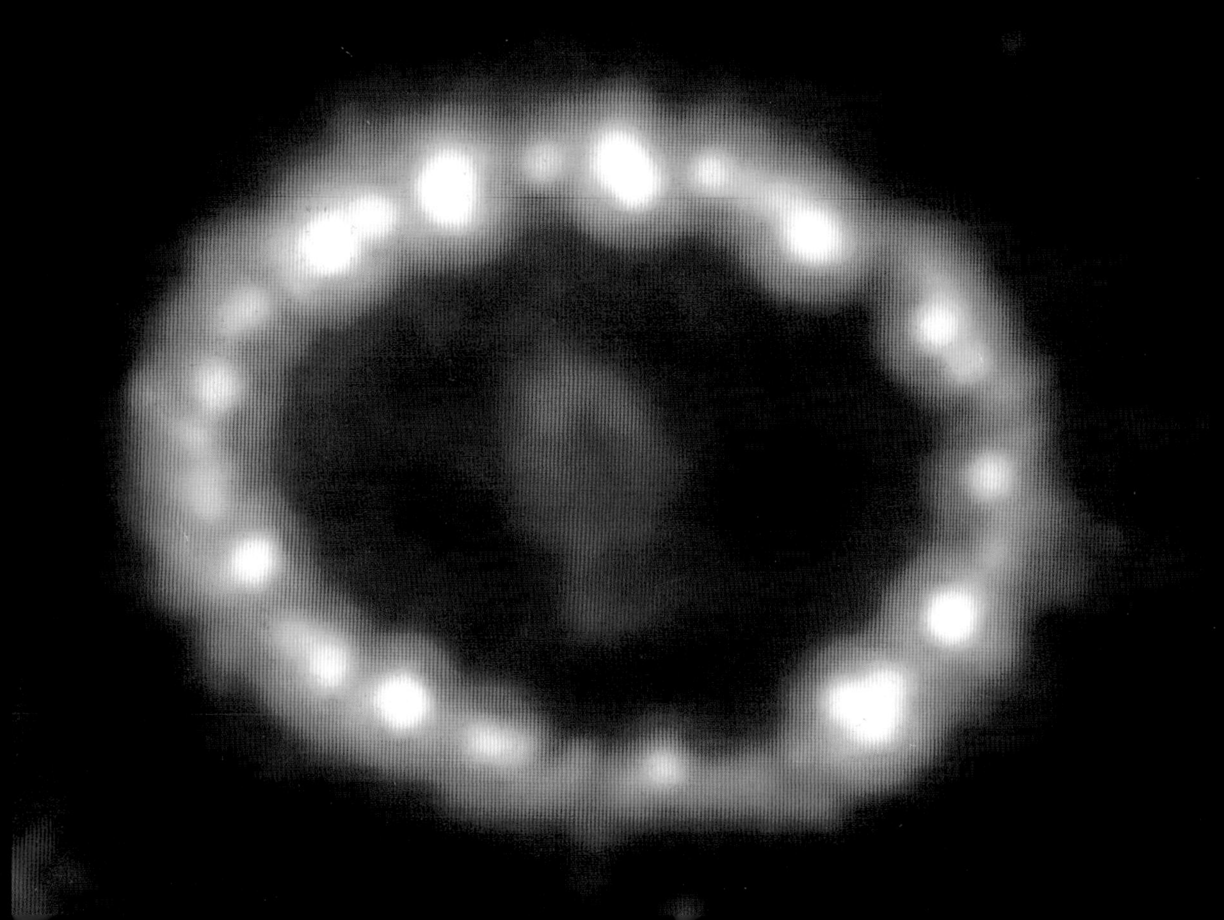

Astronomers initially assumed the remains of SN 1987A would form a neutron star. However, they have yet to find it among the remnants of the explosion. Nor is there a black hole, and that has led some to suggest a third theory: SN 1987A has become a quark star, an object so dense that even neutrons collapse under their own weight.

86 Magellan Probe

VENUS HAS ALWAYS BEEN SHROUDED IN MYSTERY THANKS TO ITS IMPENETRABLE CLOUDS. However, in 1990, the Magellan probe arrived in orbit to solve the matter once and for all.

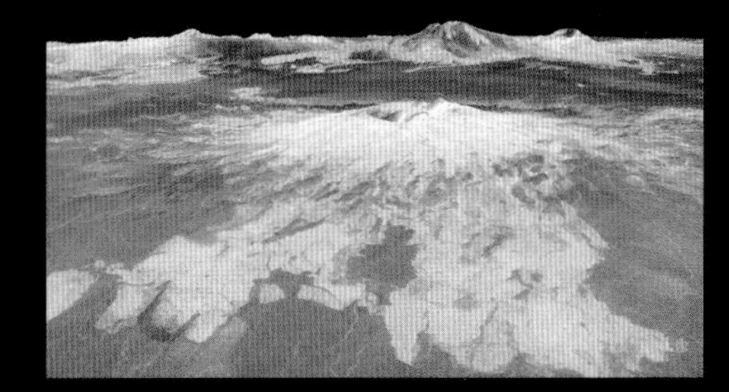

A false-color image (assigning colors to wavelength) of Sapas Mons, a Venusian volcano, compiled from the radar map made by the Magellan orbiter.

Missions to Venus have a checkered history. Early landers were crushed under the planet's enormous atmospheric pressures, and the sensitive equipment of later, tougher models had a tendency to melt in the oven-like temperatures. Magellan took a different approach. It orbited the planet for four years mapping the surface using a radar that could penetrate the blanket of caustic clouds. The surface it revealed was a volcanic hell-hole of thick lava fields. The few impact craters found suggested the entire planet was regularly resurfaced. Unlike Earth's malleable crust, Venus's rigid surface holds firm—until the pressure gets too much, and the planet is struck by tumultuous eruptions.

87 Cosmic Background Explorer

IN 1992, "RIPPLES" WERE FOUND IN THE COSMIC MICROWAVE BACKGROUND RADIATION. The result was another brick in the foundations of the Big Bang theory.

The cosmic microwave background (CMB) can be understood as the fading echoes of the Big Bang itself. It is the last remnants of that gargantuan release of energy, which can be picked up in the form of microwave radiation—short-wavelength radio waves. Found by accident in 1964 by two radioastronomers, the CMB became tangible evidence for the Big Bang theory, helping it to become the dominant idea in cosmology.

According to the theory, in the first thousandth of a second of time, the matter created by the Big Bang was annihilated by an equal amount of antimatter. However, if the two types were exactly equal, the Universe would have become an utterly empty space in a fraction of a second. The CMB ripples, discovered by the Cosmic Background Explorer probe, showed that matter was not spread evenly throughout the embryonic Universe, meaning pockets of matter were left among the emptiness, eventually to form the stars and galaxies.

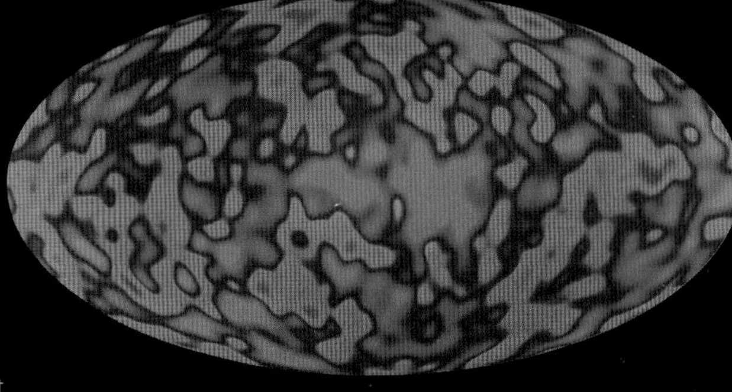

A map of the CMB shows the microwave radiation coming from all points of the sky.

88 The Hubble Space Telescope

The images produced by the Hubble Space Telescope, such as this one of the Sombrero Galaxy, have revolutionized the way we picture the Universe in the 21st century.

SNATCHING TRIUMPH FROM THE JAWS OF DISGRACE, the Hubble Space Telescope could have been the most expensive mistake in history, but has become our all-seeing eye (almost) in space.

The history of telescope astronomy has been one of building bigger and better. Lenses became smoother and clearer, before giving way to mirrors, which grew larger and larger to capture more starlight. But size was not everything, even the largest telescope needs a clear view, and the world's leading observatories began to cluster on dry mountaintops where the air was still and rarely cloudy. But light behaves differently in air than it does in the vacuum of space—causing twinkling for example—and beams of other radiation, X-rays and UV, for example, barely reach the surface. In 1923, Hermann Oberth, the German rocket pioneer, suggested that putting a telescope in orbit beyond the atmosphere would afford the clearest view of all.

Mirror, mirror

Space-based telescopes had been tried before—the Skylab had one—and it had been a goal of the NASA space program to launch one since the 1960s. However, budget cuts and accidents meant that the Hubble Space Telescope, a 12-ton metal tube, did not get into space until 1990, heaved up 559 km (347 miles) by the space shuttle Discovery.

At first all seemed fine, the 2.4-meter mirror (still dwarfed by Earth-based telescopes) was producing images clearer than had ever been seen before. However, they were not as clear as they should have been. It turned out that the mirror was the wrong shape—albeit by a few nanometers (billionths of a meter) but that was enough to make the HST images 10 times fuzzier than the mission specs demanded.

The error was studied very precisely and in 1993, shuttle astronauts were sent to fit "spectacles"—new components that would correct the error and let HST see as intended. The repair mission, a lasting testament to the brilliance of the space shuttle concept, took 10 days of precise EVA work, with the space telescope held over the cargo bay.

Seeing far, seeing early

The HST is something of a time machine. It can see further into the Universe than any other telescope, at objects billions of light-years away. That means their light has traveled for billions of years before entering the HST. The image they form shows what the sky looked like all those years ago, so HST can see the Universe when it was young. So far the HST has seen back 13 billion years, to perhaps just 500 million years (no one is quite sure) after the Big Bang itself.

The HST is a Cassegrain reflecting telescope named for its French designer whose work in 1672 was overshadowed by the Newtonian version. In a Cassegrain device the light collected by the main mirror is focused on a secondary, which reflects it straight back through a hole in the center of the primary mirror. Where the human astronomer would be, HST has electronic optics, similar to those in an everyday digital camera.

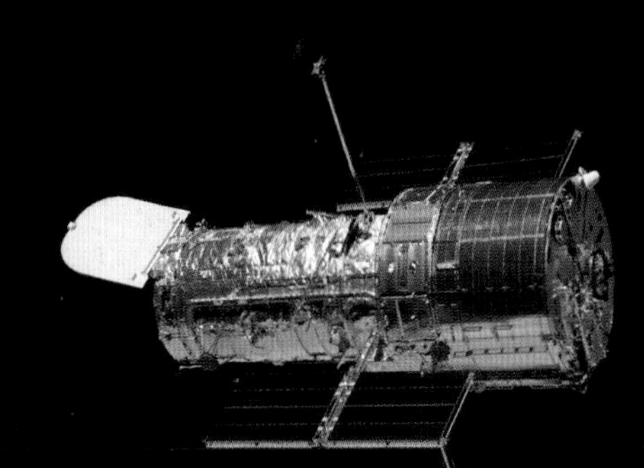

89 Comet Crash

SUCCESS IN SPACE IS ALL ABOUT PLANNING FOR EVERY EVENTUALITY. IN JULY 1994 WHEN A RATHER UNEXPECTED event was discovered, a space probe was ready and able to offer a front-row view. Astronomers were going to watch the largest impact in recorded history.

The SL9 comet hit the far side of Jupiter out of the sight of Earth, but since Jupiter rotates more than twice as fast as Earth, the impact sites were soon visible as dark regions among the banded Jovian clouds.

Thankfully the planet was not Earth, and the comet was a tenth of the size of the one that hit Earth 69 million years ago, destroying the dinosaurs. Nevertheless the comet in question was going to make a big hit. As its name suggests, Comet Shoemaker-Levy 9 was discovered by astronomical couple Carolyn and Eugene M. Shoemaker and their colleague David Levy. And SL9 was the ninth one this team had found. When it was spotted in 1993, the comet was something of an oddity—it orbited Jupiter not the Sun, the first of many to be found there. SL9 went around every two years. On its last pass in 1992, the comet had got so close, Jupiter's gravity had ripped it into 21 chunks. The next time around, these objects were going to hit the planet.

Impacts of this scale only happen every couple of centuries, and astronomers did not want to miss it. As luck would have it a probe called Galileo was already on its way to Jupiter. Controllers turned it around so its camera faced the right way. It was expected that the impact sites would churn up material from deep inside Jupiter. However, they appeared as dark bruises of hydrogen sulfide and sulfur from just below the cloud tops. The ice and rock fragments shattered on impact and did not penetrate as far into Jupiter as was expected.

GALILEO HITS JUPITER

After the spectacular comet show, Galileo continued its mission. The first thing the craft did on arrival at Jupiter was drop a smaller probe into the atmosphere. This tough little ball of detectors plunged through the clouds sending back data before being crushed into inaction. The rest of the spacecraft then entered a complicated orbit which allowed it to pay a visit to several moons and swing low over the tops of Jupiter's clouds. Galileo revolutionized the way we think about the Jupiter system—it is now the number one candidate to harbor life. In 2003, Galileo "deorbited"—burning up in Jupiter's clouds.

90 Go SOHO

THE SOLAR AND HELIOSPHERIC OBSERVATORY, SOHO FOR SHORT, IS AN EYE IN THE SKY DOING WHAT NO HUMAN CAN—LOOK DIRECTLY AT THE SUN. Launched in 1995, SOHO has seen its mission extended seven times. With plenty of power from its solar panels, it may work for many years to come.

Unlike the planets, comets, or asteroids that are targeted by space probes, the Sun occupies a stationary point. That widens the launch window for solar probes considerably and provides options on how best to study the Sun. NASA's Pioneer program of the 1960s created a constellation of probes, with four spacecraft in solar orbit at roughly the same distance as Earth, but at different angles. The Pioneers monitored "solar weather," such as the strength of the solar wind and magnetic field. In the early 1990s, Ulysses, a European probe, flew over the Sun's poles and discovered that the corona, the Sun's halo of plasma, was missing there.

SOHO carries 12 detectors, most of which also have easily pronounced acronyms: SWAN, GOLF, VIRGO. Most are looking at the corona and the wider heliosphere (the region filled with the solar wind) using ultraviolet telescopes and are also detecting internal changes by measuring the solar seismic activity—the rise and fall of the Sun's surface. SOHO has given the best view in a halo orbit of the first Sun–Earth Lagrangian point, which is about 1.5 million km (932,000 miles) from Earth toward the Sun. That means it goes around the point in space where the gravitational pull of the distant Sun is equal and opposite to the pull of the much nearer Earth, orbiting in a plane perpendicular to the Earth, hence the halo.

SOHO lost contact with Earth in 1998 for two months after its gyroscopes failed. Powerful space radars were used to locate it and a new control program started it working again.

91 Aliens Found?

THE IDEA THAT MARS IS HOME TO EXTRATERRESTRIAL LIFE HAS BEEN POPULAR FOR MORE THAN 150 YEARS. Martians are imagined as little and green, belligerent invaders, or a guardian race watching over us. It is little wonder that the first scientific reports of alien life came from Martian specimens, albeit rather unimpressive ones.

The presence of apparent biomorphs in the Martian rock prompted U.S. President Bill Clinton to make a televised address. Since then skeptics have argued that the bugs are too small (they are less than 100 nm) to contain RNA or are contaminants from Earth.

No mission to Mars has yet brought back a sample of the planet's rock. But that does not mean that we do not have any samples from our smaller neighbor. Just like Earth and the other rocky planets, Mars has had its fair share of meteorite strikes. The bigger ones hit with such violence that they flung chunks of Mars into space. Some of these dispossessed bits of Mars found their way to Earth, raining down as meteorites themselves. So, space probes may not have sent rocks back from Mars, but little bits of the planet have made it to Earth. Meteorites are hitting Earth all the time—perhaps two a day are large enough to make it all the way to the ground without burning up. Most are never found, and are lost among Earth's own rocks. One place where it is easier to spot meteors is Antarctica, where the dark space rocks stand out among the white ice. In 1984, one such polar meteorite hunt found rock number ALH84001, which turned out to be a piece of Mars.

ALH84001 is code for rock number 1 from the Alan Hills of Antarctica found in 1984. It weighed just under 2 kg (4.4lbs) and is thought to be a little over 4 billion years old. The latest theory suggested that the rock hails from the rugged Valles Marinensis region of Mars where it was formed by a meteorite impact. It was propelled into space by another meteorite strike 15 million years ago and then hit Earth 13,000 years ago.

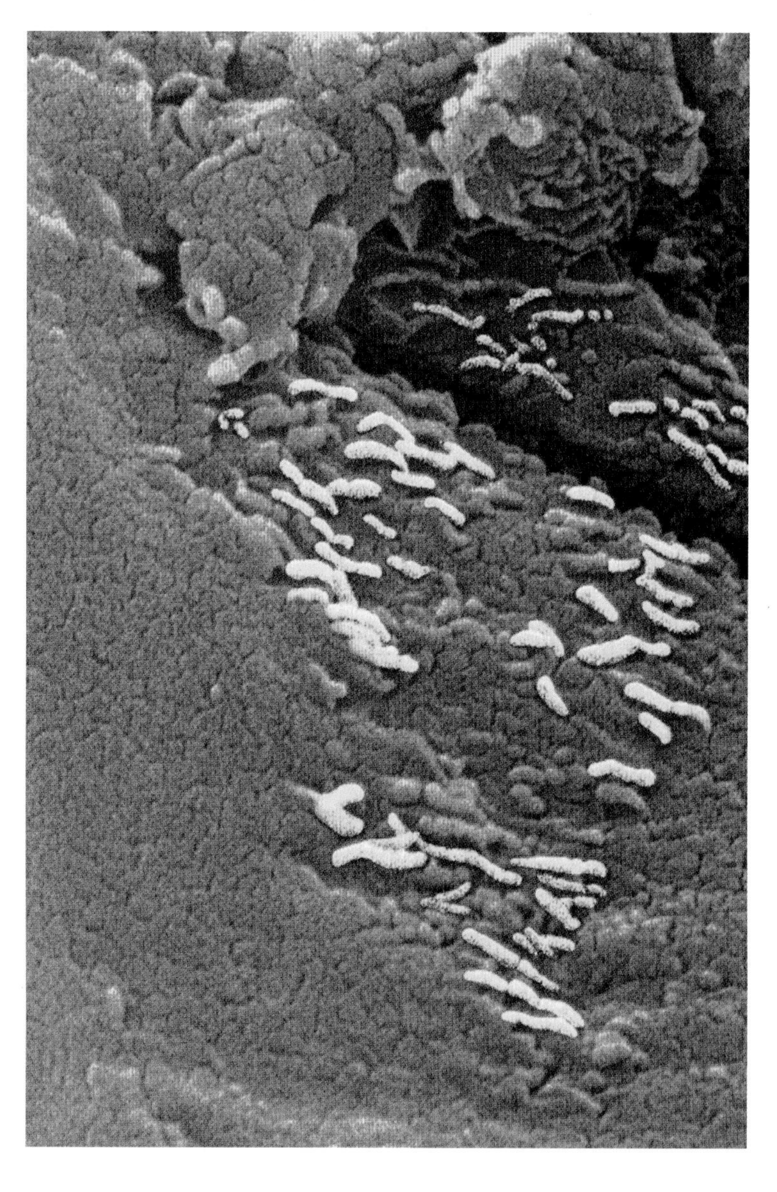

Closer look

In 1996, NASA scientists took a look at the rock under an electron microscope and saw what looked very much like fossilized bacteria. It was announced that this was evidence of life on Mars in the distant past. Not everyone was convinced about these miniature Martians but the discovery ensured that NASA was again asked to send probes to Mars to look for better evidence of life.

92 Dark Energy

FROM NEWTON THROUGH EINSTEIN TO HUBBLE, IDEAS OF AN EXPANDING UNIVERSE AGREED ON ONE SIMPLE RULE—AS THE UNIVERSE GREW AND GOT OLDER the rate at which it expanded would slow. Then in 1998 a star survey obscured that understanding of space and time—and we are still in the dark.

For 70 years Hubble's law ruled: every distant galaxy cluster in the Universe is moving away from all the others. We all knew they were thrown apart by some explosive event in the distant past, a Big Bang, let's say. But Newton's centuries-old contribution was still valid: gravity attracts masses, so the received wisdom was that the expansion of the Universe was surely slowing under the drag of all that gravity pulling galaxies back toward each other. The questions then were would gravity one day stop the expansion, would it precipitate a contraction, or would the power of the Big Bang outstrip the pull of gravity leading to eternal expansion?

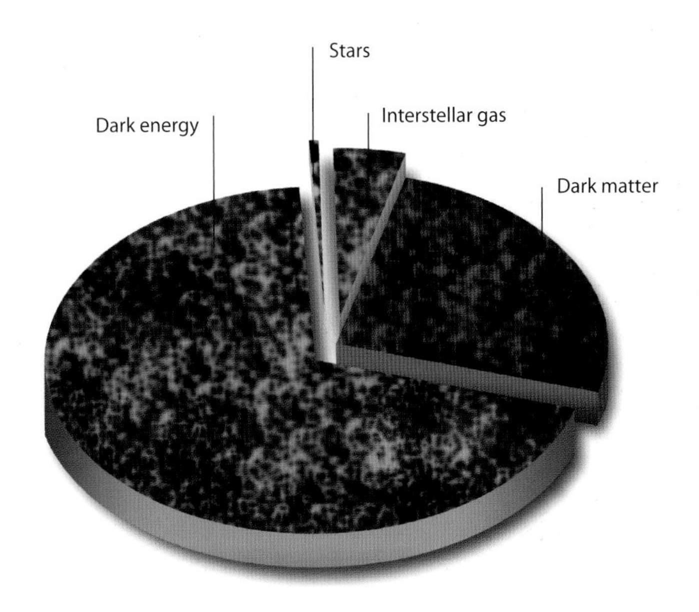

The increasing rate of expansion now measured for the Universe suggests the atoms, planets, and stars are an even less significant portion of the Universe than had been thought. Visible matter is less than one per cent of the whole. Add in around three per cent for the gases and dust that is the same kind of matter (just hard to see), that leaves the rest as unexplained with dark matter at 22 per cent and dark energy at 74 per cent.

Gravity is proportional to mass, and by now it was understood that most of the Universe's matter was invisible, or dark. It was decided that measuring how fast the expansion was slowing would throw light on this mysterious stuff. A survey of type 1a supernovae was begun. These were formed by binary systems of a main sequence star and white dwarf. Material from the larger neighbor is pulled over to the white dwarf until its mass reaches the Chandrasekhar limit—whereupon it is too large to support itself and, boom, it goes supernova. Type 1a stars are all the same mass and magnitude and so can be used as standard candles to measure distances across the Universe—the fainter ones are farther away. The redshifts show how fast an object is moving away, and so it was expected that the more distant (and older) light would show a greater redshift than the nearer (and younger) sources. Older light would show a past expansion rate which could be compared with the rate now.

The astronomical community was shocked by the result of the supernova survey. The expansion of the universe is not slowing; gravity is not pulling everything to a halt. In fact expansion is speeding up! The as yet unfathomable entity that is causing this acceleration has been named dark energy. No one has measured dark energy, only its effects. Theories on what dark energy is are beginning to link the immense nothingness of spacetime with energy and mass in their smallest possible quantities. Even nothing has some energy, it appears, and there is a lot of nothing out there.

93 The World in Space

THE THAW IN WORLD POLITICS THAT CONTINUED THROUGH THE 1990S WAS REFLECTED BY INCREASING COOPERATION IN SPACE PROJECTS. This reached its zenith with the International Space Station (ISS), launched in 1998 and now the largest spacecraft in history, playing host to space travelers from 15 countries and even a few paying guests.

The huge solar arrays of the ISS make it about the same size as an American football field. At certain times of year, when it catches the low sunlight, it is possible to see the spacecraft with the naked eye as it moves at a noticeable speed across the sky.

The success of the Soviet Mir showed a way that space agencies could clock up more hours in space on lower budgets. Launching heavy loads into space was the costliest part of space travel. But, once built, a space station could house a crew for long periods. Only relatively small rockets were needed to send up supplies and replacement astronauts.

The International Space Station was born partly because the Russian Space Agency (restyled from its Soviet heyday), was not able to afford Mir 2. Meanwhile NASA decided to shelve their own station, named Freedom, and go into partnership with their new Russian friends. By 1998, the Japanese and Europeans had also joined forces, ready to contribute plans from their proposed space stations. The Canadian space agency, famed for its space robotics, was involved too when the first module Zarya, ("dawn" in Russian) went up.

By 2000, as Mir was abandoned, soon to end its 15 years in space by crashing into the South Pacific, the ISS got its first permanent crew. It has been occupied ever since, gradually enlarged by its owners so that it now has 12 pressurized modules, including labs, crew cabins, and a cupola for a bit of stargazing. Since the retirement of the NASA shuttle, and in lieu of an American crew vehicle, the ISS is serviced entirely from the Baikonur Cosmodrome in the desert of Kazakhstan. Life on the space station has become routine with a new breed of scientist astronauts replacing the fighter pilots of early days. In 2001, Denis Tito, an American finance magnate, paid $20 million to spend 8 days in the Russian quarters of the ISS—the first space tourist in history.

94 Is Earth Special?

AS FAR AS WE KNOW EARTH IS THE ONLY PLACE IN THE UNIVERSE THAT HARBORS LIFE. THE CONDITIONS NEEDED FOR LIFE ARE WELL UNDERSTOOD AND MANY ASTRONOMERS BELIEVE extraterrestrial life is a near certainty. However, in 2000 an Earth scientist and space scientist joined forces to suggest that our living planet is in fact very rare, perhaps unique.

By the 1930s it had become obvious that the Solar System was nowhere near the center of the Milky Way, and the principle of mediocrity took hold. This states that our Sun, Solar System, and its life-bearing planet are nothing unusual. All life needs are the laws of physics and chemistry plus the right conditions for a primordial soup to cook up some biochemical entities that grow and reproduce. What is needed is often called a Goldilocks orbit—a place in space that is not too hot, not too cold, but just right. Earth occupies this orbit, and is the only place known to have liquid water on its surface—and the life that comes with it.

The last mass extinction occurred on Earth 69 million years ago when a 10-km (6-mile) meteorite hit what is now Mexico. If such an event had occurred at any point since then, our civilization would have been over before it even began.

Easy come, easy go

However, in 2000, geologist Peter Ward and astrobiologist Donald E. Brownlee suggested that Earth was a rare planet after all. They were not disputing that life could start elsewhere; what was unusual, they said, was that it had not come to an end yet on Earth. Nearby gamma-ray bursts or comet strikes could wipe life from its surface, and less catastrophic strikes could frequently disrupt a planet, causing mass extinctions. It took more than the Goldilocks orbit for Earth to have kept such catastrophes to a minimum for the 3.5 billion years or more that life has been evolving. That more or less uninterrupted evolution had resulted in a civilization capable of questioning its place in the Universe. Perhaps that is what makes Earth special, even unique?

What are Earth's saving graces? The gravity from our large moon warms the Earth's interior and boosts the magnetic field which wards off nasty cosmic rays, while the pull of Jupiter, our big neighbor out toward deep space (but not too close), probably sweeps up many of the wayward comets that might otherwise smash into Earth with an alarming regularity.

95 NEAR Shoemaker

ANOTHER FIRST IN SPACE EXPLORATION WAS CHALKED UP IN 2001, WHEN A LITTLE SPACECRAFT LANDED ON AN ASTEROID, one of those orbiting rocks too small to be a planet but definitely large enough to get noticed. Of extra interest are the NEAs, the near-Earth asteroids that might one day hit us.

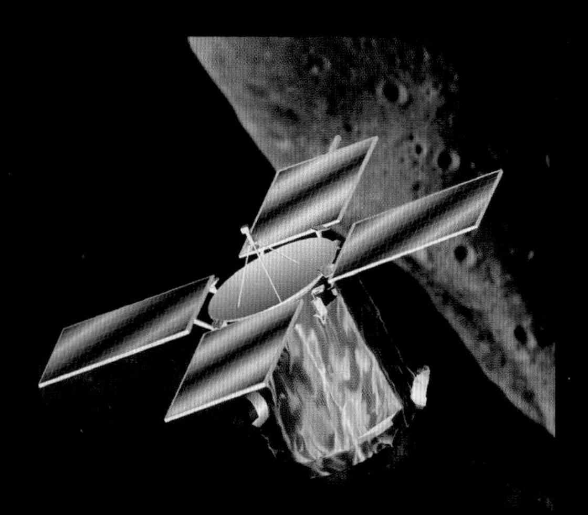

Cleverly named NEAR (Near Earth Asteroid Rendezvous) on launch, Shoemaker was added to the name mid-flight to honor Eugene Shoemaker, the great astrogeologist (expert in space and rocks) who had just passed away. The little probe was heading for Eros, a 34-km-long (21.1 miles) peanut-shaped rock that orbits the Sun and frequently comes close to Earth. NEAR missed its first orbit around Eros due to engine problems and had to fly right around the Sun for another go in early 2000. In 2001, it landed in the narrow middle section of the asteroid, where it sent back data for 14 days. The probe found that Eros's gravity was stronger at its bulbous tips. Potentially that means small grains on its surface could roll uphill!

NEAR Shoemaker traveled 3 billion km (1.86 billion miles) to land on Eros. Astronomers were keen to get there because it is possible that Eros's orbit may change to cross that of Earth—and then it will be just a matter of time before both asteroid and planet find themselves in the same place at the same time.

96 The Oort Cloud and Kuiper Belt

ONE OF THE PUZZLES OF ASTRONOMY WAS WHERE ALL THE COMETS CAME FROM. OVER BILLIONS OF YEARS, countless comets had crashed into anything and everything, including Earth, but they did not appear to be reducing in number. The answer was found at the edge of our Solar System.

The Solar System formed from a disc of dust and gas that was left over when the Sun grew large enough to ignite. The dense materials, like metal and rocky minerals, were drawn by gravity towards the inner region, where they made the four rocky planets. Further out, the outer planets were formed from low-density gas and ice (it was colder out there). Comets were "dirty snowballs," lumps of ice and rock left over from the dawn of the Solar System, 4.6 billion years ago. In the 1950s, the eminent Dutch astronomer Jan Oort suggested that comets were visitors from a distant cloud of icy lumps that lay near the edge of the Solar System, where they had been too far

DUSTING FOR CLUES

In 2005, the Deep Impact probe (below) was sent to fire a missile at a short-period comet, Tempel 1. The copper spike created a cloud of dust and a crater that was observed by the probe's detectors. The year before, the Stardust probe had swept some dust from the coma of Wild 2, another comet. The dust, encased in gel, was parachuted back to Earth in 2011. Both missions showed a slush of waxy ices and clay-like dust.

away to become involved in the formation of planets. He thought comets were knocked from the cloud by the gravity of a passing star many light years away disrupting their fragile equilibrium bodies. However, comet hunters pointed to differences in the orbital periods of comets—the time it takes for them to circle the Sun. Short-period comets take fewer than 200 years; long-period comets can take thousands. While the trajectories of the long-period objects tallied with an origin in what had become known as the Oort Cloud, 1,000 times further out than Pluto, the paths of the short-period comets suggested they came from closer to home.

Many Planet Xs

Since the days of the Planet X theory, when astronomers thought there was a large object beyond Neptune (the search for it found Pluto), few people had looked into the Solar System beyond the planets. Maybe there were smaller objects out there, and perhaps this was where the short-period comets were coming from.

A survey in the late 1980s searching manually for these hypothetical bodies employed the same blink comparator technique that had been used to find Pluto 60 years before. This was soon automated using CCDs, electronic optical kit now found everywhere in digital cameras but at the time pretty cutting edge technology. Gradually objects were discovered in what became known as the Kuiper Belt, named for another stargazing Dutchman, Gerard Kuiper (rhymes with *wiper*). In the 1950s, he had suggested a disc of this kind would be created during the formation of the Solar System, although he was of the opinion it had been dispersed long ago by Pluto—at the time thought to be Earth-sized.

KBOs threaten Pluto's status?

Triton, a large ice moon of Neptune, orbits in the opposite direction to the rest of the planet's satellites. That led to suggestions that Triton is a large KBO (Kuiper Belt object), captured by the giant planet long ago. Triton is bigger than Pluto—as is our Moon—and by 2002, KBOs were being found that were around the same size as Pluto (it is hard to measure them exactly). What would these big interlopers mean for Pluto's hallowed status as the ninth planet?

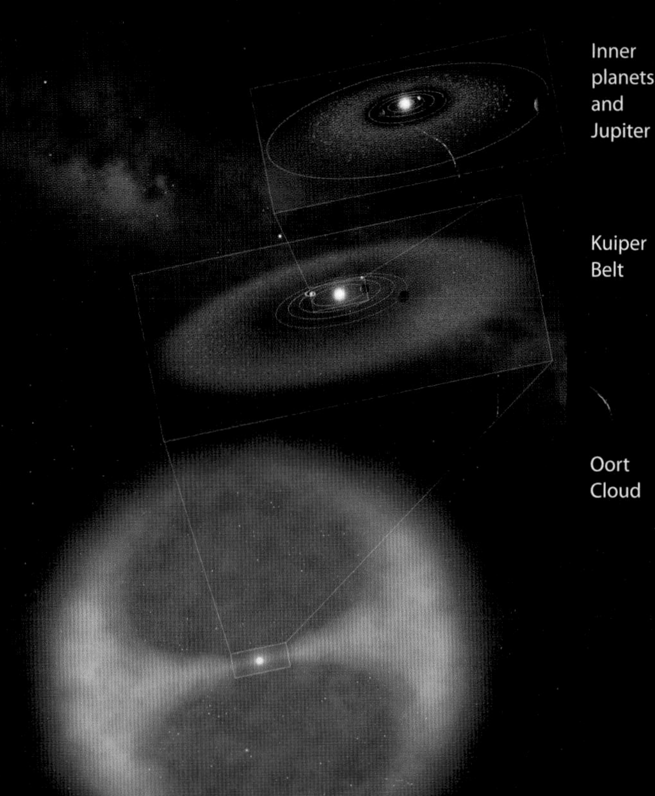

There is a lot more to the Solar System than a star, some planets, and an asteroid belt. In fact these fill just a small space in the middle. The current thinking is that the Kuiper Belt is a central disc that connects to the Oort Cloud.

97 Sending Rovers

THE SCENES OF ASTRONAUTS DRIVING AROUND THE MOON IN THE LUNAR ROVER, THE ULTIMATE DUNE BUGGY, ARE SOME OF THE MOST ENDURING IMAGES OF THE APOLLO PROGRAM. The idea of sending robotic rovers to explore other worlds was put into action in those early years but no one knew just how tough it was going to be.

The Mars Exploration Rovers—Spirit and Opportunity—arrived by parachute, and retrorockets slowed the descent before the air bags were dropped the final 10 meters (11 yards). The impact speeds were 100 kilometers per hour (62mph), making the lander bounce a dozen times and roll 900 meters (nearly 1,000 yards).

The first mobile, robotic probe to roll across the surface of another heavenly body was Lunokhod 1, an eight-wheeled rover that was the main Soviet contribution to lunar exploration. The 230-cm (90.5-inch) machine, which had a rather make-do-and-mend appearance and resembled a bathtub on wheels, arrived at the Sea of Rains in 1970. It spent the next 10 months rolling several thousand meters/yards across the lunar surface, analyzing soil samples and sending back pictures. Its batteries were charged by solar cells, and during the lunar night it shut down apart from a heating system powered by a radioactive source.

This success led to smaller skidoo-like rovers being included in the first Soviet Mars landers sent in 1971. They both crashed, then after a successful deployment of a second Russian lunar rover in 1973, the Soviet rover program went quiet.

Taking a sojourn

NASA also recognized that a rover was the best way to investigate the red planet, but it would be another 25 years before one made it down in one piece. In 1997, the Mars Pathfinder mission delivered a small rover, called Sojourner, to the red planet. A kind of solar-powered skateboard, this six-wheeled little vehicle arrived inside a bundle of airbags to break its fall. Driving Sojourner from Mission Control was a slow and deliberate job—it took 10 minutes for every command to reach the rover. In 83 days, Sojourner traveled about 100 meters (109 yards), sending back the best pictures yet of Mars and analyzing its soil for tell-tale signs of biogenic materials—anything that might have been made by alien life, extant or extinct. Soon more probes were on their way—but getting to Mars is not easy.

THE SEARCH FOR WATER

In 2004, NASA began planning to send astronauts back to the Moon, and the same spacecraft were to be used to fly to Mars by 2050. Project Constellation is no more but while it lasted a Mars probe found a crucial piece in the flight plan to the planet: water. Astronauts would have to spend several months on the surface of Mars, and a supply of water in the planet's rocks would be very useful. In 2008, Phoenix landed at Mars's north pole, where it dug a hole in the frozen ground. Phoenix found what looked like lumps of water ice (bottom left, in shadow) in the dusty soil. It was confirmed as such when four days later, the ice had melted away.

Spirit and Opportunity

The next three Martian missions crashed, so tensions were high in late 2003, when a much larger rover called Spirit landed on Mars. A few weeks later—now 2004—an identical rover called Opportunity arrived. Both were targeted at flat regions, ideal for bouncing down in airbags and with few obstacles for the landers to bump into. Each rover was cocooned in a protective pyramid, which folded open—rather ceremoniously—forming ramps down to the rust-colored regolith.

Both Spirit and Opportunity were unprecedented successes. They carried the equipment for studying the rocks on Mars and also had shovels and drills for taking soil and rock samples. Their stereo cameras scanned the landscape like a pair of eyes to measure distances and produce the incredible panoramas of the Martian desert.

Spirit rolled into deep sand in 2009 and was unable to power itself free so the rover became a static research station. However, it did not survive the 2010 Martian winter, during which the rover was supposed to park on sunlit hills and go to sleep until summer. Spirit was unable to do this and lost all power in 2011. Opportunity, however, is still going strong.

If the curse of Mars does not strike again—interplanetary exploration is far from fullproof—the next generation of rover, NASA's Mars Science Laboratory, will soon join Opportunity. MSL is the size of a station wagon. Perhaps the next rover will be one driven by a human.

Opportunity is one of the MERs, Mars Exploration Rovers, along with Spirit. It is about the size of a golf buggy, but even slower. It relies entirely on solar power and could potentially work indefinitely, but eventually the cruel winters on Mars will wear it down.

98 Landing on Titan

THE FIRST DESCRIPTIONS OF SATURN'S RINGS AND MOONS WERE MADE BY DUTCHMAN CHRISTIAAN HUYGENS and French astronomer Giovanni Cassini. In 1997 Huygens and Cassini set off for Saturn, this time robotic probes sent to survey the rings and make the first landing in the outer planetary system.

The Huygens lander, the first to touch down on a moon of another planet, fell into a muddy area strewn with rocks, and sent back pictures and atmospheric data from Titan for another 90 minutes.

The nuclear-powered spacecraft, a joint venture between NASA and the ESA, was named Cassini, while the little lander aboard was Huygens. The mission took the scenic route, circling Venus twice to get a gravity slingshot past Jupiter before arriving at Saturn in 2004. There it passed between the rings, which are only a few meters/yards thick but more than double the width of the planet. Cassini took a tour of the moons, and in 2005, dropped Huygens through the thick orange clouds of the moon Titan, Saturn's largest moon. The lander found a world covered in frozen methane, and oceans of propane and other more complex hydrocarbons.

99 Dwarf Planets

IN 2006, THE NUMBER OF PLANETS IN THE SOLAR SYSTEM WENT FROM NINE TO EIGHT AS PLUTO WAS DEMOTED TO A DWARF PLANET. A discovery the year before showed that Pluto was by no means the biggest little object in the area.

In 1978, the mass that had been thought to be just Pluto—estimated back then as little more than that of Mercury—turned out to be shared between Pluto and a very large moon. That satellite was named Charon (for the boatman who took souls to Pluto's underworld) and was about a third of the size of its "planet." Some suggested that Pluto-Charon should be regarded as a binary planet, still out there in ninth place. However as the sheer extent of the Kuiper Belt and the many larger objects within it became established, a rethink was needed.

Once Charon was on the scene it became apparent that Pluto—pictured here with Charon between Earth and our Moon—was not quite the body everyone had been imagining since 1930. In 2005, Pluto was found to have two other tiny moons named Nix and Hydra, plus a fourth, as yet unnamed, was spotted in 2011.

Following the discovery of Eris, a KBO that was slightly larger than Pluto in 2005, the International Astronomical Union decided to act the following year. Pluto, Eris, two other KBOs, Haumea and Makemake, and the largest asteroid Ceres were to be retermed dwarf planets. What set them apart was that firstly these objects, like planets, were large enough for their gravity to pull them into a sphere (although Haumea is more like an egg). Secondly, and unlike a planet, they were not large enough to clear their region of other objects. As the planets formed they drew all nearby material to them, perturbing any other objects so they crashed into the young planet. As a result, planets orbit in empty space. Not so the dwarfs, with Ceres surrounded by asteroids and the others in the Kuiper Belt. So far there are just five dwarf planets but many more are currently being considered. Within a few decades there will be dozens of dwarf planets on the books, all needing names. Currently the newest discoveries are still named after deities as tradition dictates, just not Greco-Roman ones.

100 A New Earth

A NEW SPACE TELESCOPE NAMED FOR THE ASTRONOMER WHO DISCOVERED HOW PLANETS ORBIT WAS LAUNCHED IN 2009. Kepler had one goal: to survey the Milky Way, find stars that had planets, and ascertain if any were in a habitable zone. The search for Earth no. 2 had begun.

Astronomers had confirmed a sighting of the first "exoplanets" in 1992. These were truly alien worlds, small lumps of rock that orbited a pulsating neutron star, and nothing like our solar system. The search for habitable planets continued. It is not as simple as watching a star and waiting for a planet to appear in orbit. Astronomers use a number of other detection methods. Some look for the wobble in the motion of a star caused by planets moving around it. The wobble is seen in tiny shifts in its spectra. Others wait for stars to dim as their planets move in front, relative to Earth, obscuring starlight slightly.

Within four months—although it took many more to process the data—Kepler had seen more than 1,200 objects that looked like planets. By 2011, 68 of them were thought to be near the size of Earth with five appearing to orbit the habitable zone, where liquid water would exist. However, there is a lot more work to do to confirm that they are anything like Earth.

The latest estimates suggest that the number of planets in the Universe is actually higher than the number of stars. Even if the chance of an advanced civilization is just one in a billion, there could very well be more than 100 civilizations in the Milky Way alone.

The Kepler telescope uses the transit detection method, looking for changes in starlight as planets move around their stars. Beyond the atmosphere, free of twinkles and other distortions, it can detect the reduction in light from a star produced by a planet up to 100 times smaller than Earth.

101 Astronomy: **the basics**

SO WHAT DOES ALL THIS DISCOVERY ADD UP TO? If we take a look at astronomy from another angle, drawing all the lines of inquiry together we can explore the very fundamentals of the subject.

The four forces

A force is a transfer of energy from one mass to another and has the effect of altering the motion of both bodies. Over the years physicists have isolated four fundamental forces at work on Earth. The basic tenet of astronomy is that the laws of physics as observed on Earth also apply in all other corners of the Universe. Therefore the properties of the four forces underlie our understanding of how stars form and how planets move, and allow us to interpret the lights we see in the sky.

The first force is the strong interaction. This holds together the protons and neutrons that form the core, or nucleus, of an atom. It is the strongest of the lot, as

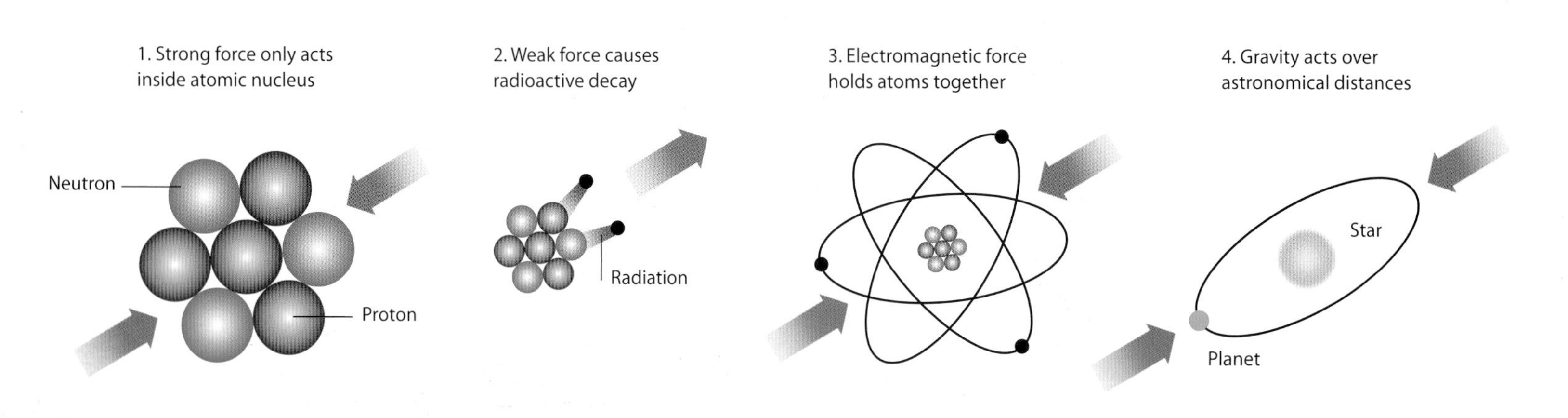

its name suggests, but its influence barely extends beyond the nucleus. Force two, the weak interaction, is involved in radioactivity, where particles are pushed out of an unstable atomic nucleus. (The ejected material is detected as radioactivity, or nuclear radiation.) Thirdly we have the electromagnetic force, which is the "opposites attract, likes repel" force. It holds the negatively charged electrons in place around the positively charged nucleus. The same force stops two atoms merging. The negative electrons of each repel each other—giving matter its form and resistance to change. Electromagnetism is also at work on a larger scale in magnetism and electricity. Finally, force four is gravity. This is a force of attraction that acts between anything with mass. A larger mass pulls more strongly on the smaller one, and the gravity of trillions upon trillions of bodies is what sets all objects on their respective paths through space.

Observing radiation

Astronomy is an indirect science. It is not possible to sample most of the Universe—it is all too far away. Instead astronomers gather information by collecting light and other radiation from the sky. Only a small part of the radiation from space is visible to the eye. Earth's atmosphere is transparent to visible light—that is why we have evolved to see by it—and radio waves pass through air with little difficulty. However, most of the infrared from stars, UV in the sky, and high-energy X rays and gamma rays do not reach the surface. Telescopes get the best view of these rays outside the atmosphere in orbit.

EYES TO THE HEAVENS

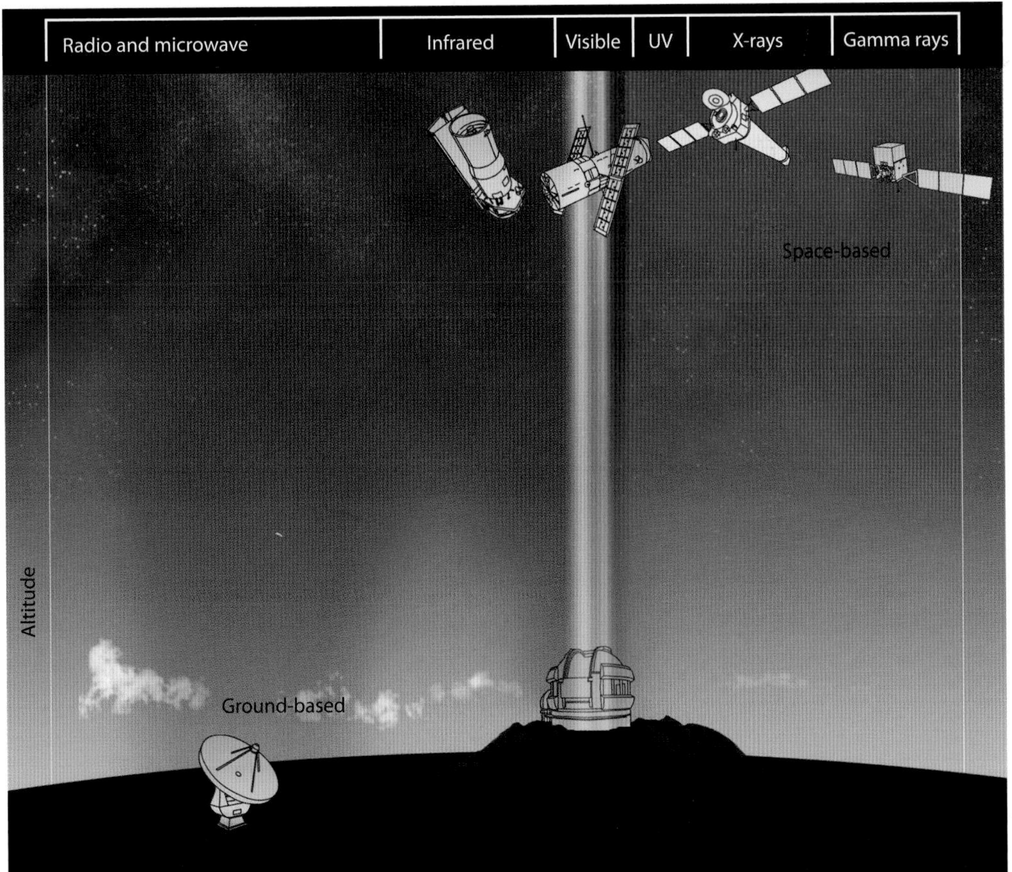

MOON
Earth's only satellite is the largest in the Solar System when compared to the size of its parent planet. It is roughly a quarter as wide as Earth at 3,475 km (2,159 miles).

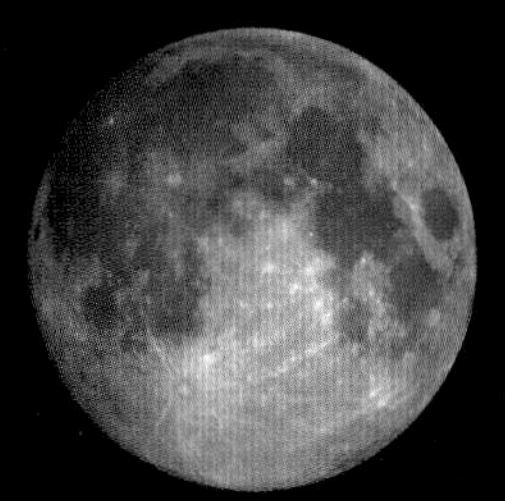

SUN
A yellow dwarf star about half way through its 10 billion year life. Toward the end the Sun will swell up to a red giant that will engulf Mercury and Venus, and make Earth the first planet. Earth's atmosphere will be blasted away ending the possibility of life. However, the moons of Jupiter will warm and may form a haven for whatever intelligent life remains on Earth 50 million centuries hence.

Diameter: 1,392,000 km
864,949 miles
Surface temperature: 5,500°C
Core temperature: 15 million°C.

MERCURY
The first planet has a large metallic core and thin crust. The powerful solar wind has more or less stripped away all of the planet's atmosphere.

Diameter: 4,878 km
3,031 miles
Distance from the Sun: 0.4 AU
Year: 88 days
Day: 58 days
Moons: 0
Surface temperature: 427°C

VENUS
The hottest place in the Solar System, due to the greenhouse effect of its thick atmosphere of carbon dioxide and sulfurous gas. Venus rotates in the opposite direction to most of the other planets, and the Venusian year is shorter than its day.

Diameter: 12,104 km
7,521 miles
Distance from the Sun: 0.7 AU
Year: 225 days
Day: 243 days
No. of moons: 0
Surface temperature: 460°C

EARTH
The largest of the rocky planets, Earth is the only place to have liquid water, and an ocean covers 70 per cent of it at an average depth of 4,200 meters.

Diameter: 12,756 km
7,926 miles
Distance from the Sun: 1 AU
Year: 365 days
Day: 24 hours
No. of moons: 1
Surface temperature: 14°C

MARS
A cold desert planet, Mars is likely to be the first one visited by human explorers. It is near to the Asteroid Belt, and its two moons, Phobos and Deimos, are large asteroids captured by the planet's gravity.

Diameter: 6,787 km
4,217 miles
Distance from the Sun: 1.5 AU
Year: 678 days
Day: 24.5 hours
No. of moons: 2
Surface temperature: -20°C

The Solar System

The Solar System formed 4.5 billion years ago when a supernova in the Orion Arm of the Milky Way sent a shockwave into a cloud of hydrogen, helium, and small amounts of other elements. That disruption caused the cloud to contract under its own gravity, forming a spinning ball of hydrogen that grew large enough to ignite as a new star.

The rotation spread a disc of left-over dust, ice, and gas around the star. The contents of this disc were constantly colliding, gradually clumping together into larger bodies, embryonic planets termed planetesimals. Planetesimals grew and grew, as their gravity swept up smaller objects in the region. Inner planetesimals were made from dense metals and rocky minerals; outer ones were low-density gases and slushy ices. After a tumultuous period of collisions that lasted 500 million years, the new solar system settled down with the eight planets—four small and rocky, four giants made from gases, liquids, and ices. All that was needed was a civilization to claim this patch of space.

SATURN

The second largest but lowest-density planet. Composed mainly of gas, Saturn would float in water if you had a large enough bucket to put it in. The rings are either the remains of a smashed ice moon or a protolunar disc of material that was prevented from forming a moon in the first place by Saturn's tidal forces.

Diameter: 120,540 km
74,900 miles
Distance from the Sun: 9.6 AU
Year: 29.5 years
Day: 10.5 hours
No. of moons: 48
Surface temperature: -168°C

URANUS

Somewhat dull-looking at first, Uranus is a featureless orb of methane over a slushy ice interior, devoid of activity. However, it has had its fair share of action. In the distant past something big enough to knock Uranus off its axis hit the planet. Now the ice giant does not so much spin, but "rolls" around the Sun on its side.

Diameter: 51,118 km
31,763 miles
Distance from the Sun: 19.2 AU
Year: 84 years
Day: 18 hours
No. of moons: 27
Surface temperature: -200°C

NEPTUNE

The last planet is also the windiest. With little or no weather changes to create turbulence, the same winds have been blowing around Neptune for millions of years, with nothing to stop them. The Voyager 2 probe picked up winds of 2,000 km/h (1,243 mph)!

Diameter: 49,528 km
30,775 miles
Distance from the Sun: 30 AU
Year: 169 years
Day: 19 hours
No. of moons: 13
Surface temperature: -212°C

JUPITER

The biggest planet is largely made of gas. It is likely that an Earth-sized solid core of rock and ice exists inside. The giant planets spin very fast, and Jupiter goes around fastest of all. The equatorial regions move faster than the polar ones, churning up the atmosphere.

Diameter: 142,800 km
88,732 miles
Distance from the Sun: 5.2 AU
Year: 11.9 years
Day: 10 hours
No. of moons: 63
Surface temperature: -124°C

Looking at eclipses

An eclipse is the most noticeable of astronomical events, sometimes becoming mass-participation gatherings. They can be seen without any equipment and often happen in the middle of the day. There are two types, lunar and solar. In the former, the Earth sits between the Sun and the Moon. Its shadow creeps across the lunar surface over a period of a few hours, biting a chunk from it. In a full lunar eclipse, the Moon turns red as only the twilight refracted through Earth's atmosphere illuminates its surface.

In a solar eclipse, it is the Moon that sits between the Sun and Earth, casting a cone of shadow onto our planet. This shadowy zone sweeps across the surface of Earth as the planet spins. Anyone inside the penumbra shadow can see the Moon obscuring part of the Sun. Amazingly, even a small crescent of sunlight is enough to light the world. In the umbra, or central shadow, the entire sun is obscured turning day to night for a few extraordinary seconds.

It is a coincidence that the positions of the Moon and Earth match their relative sizes and make eclipses possible.

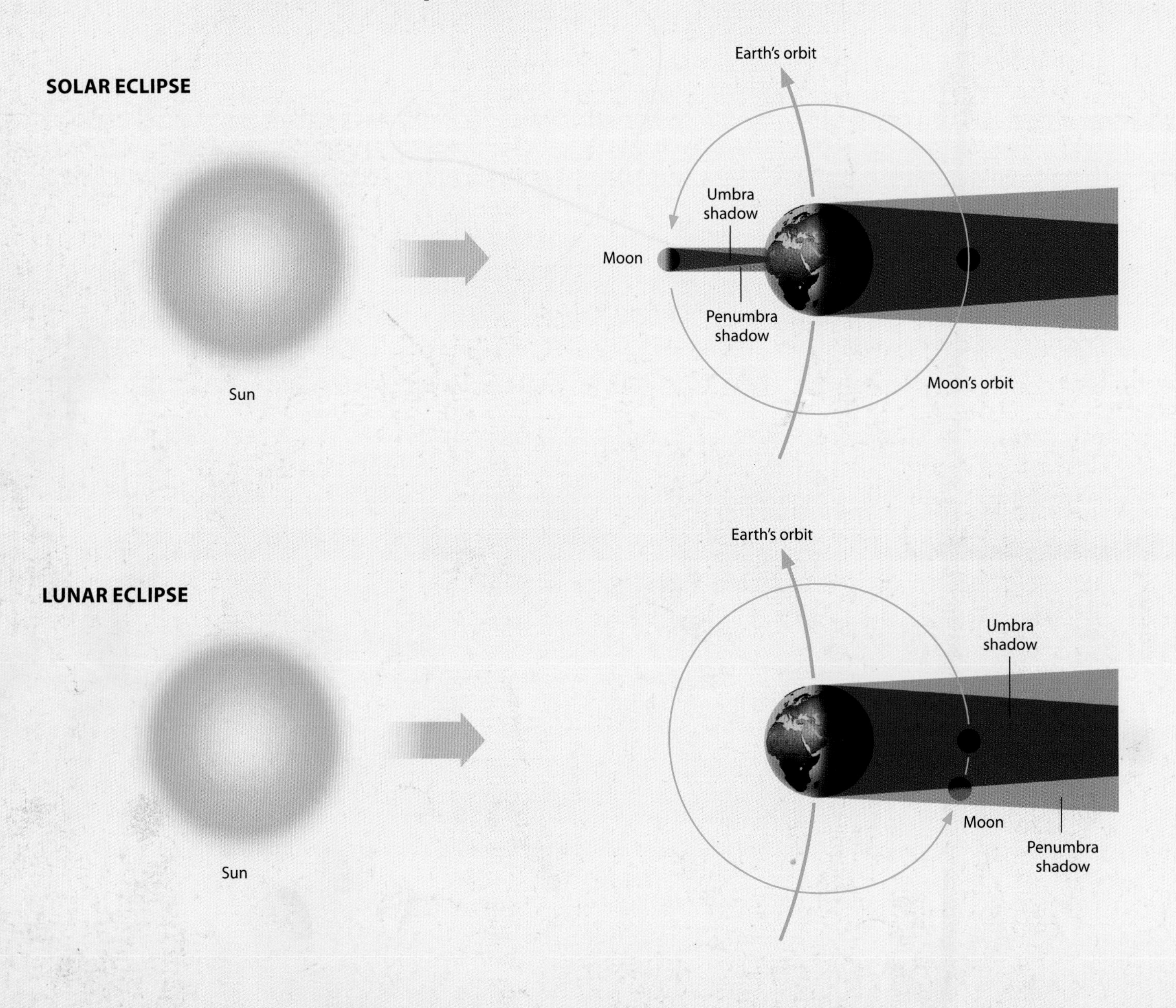

The life and death of stars

All stars are born in an immense cloud of gas, mostly hydrogen left over from the Big Bang but also some other materials released from older, dead stars. Gravity pulls some of the gas into an ever tighter ball until the center is under a pressure high enough for nuclear fusion to start. The energy that is released is what makes the star shine—emit light, heat, and other radiation. The hydrogen is the star's fuel and once it is all gone the star will begin to die. The way this happens depends on the size of the original star.

STELLAR EVOLUTION

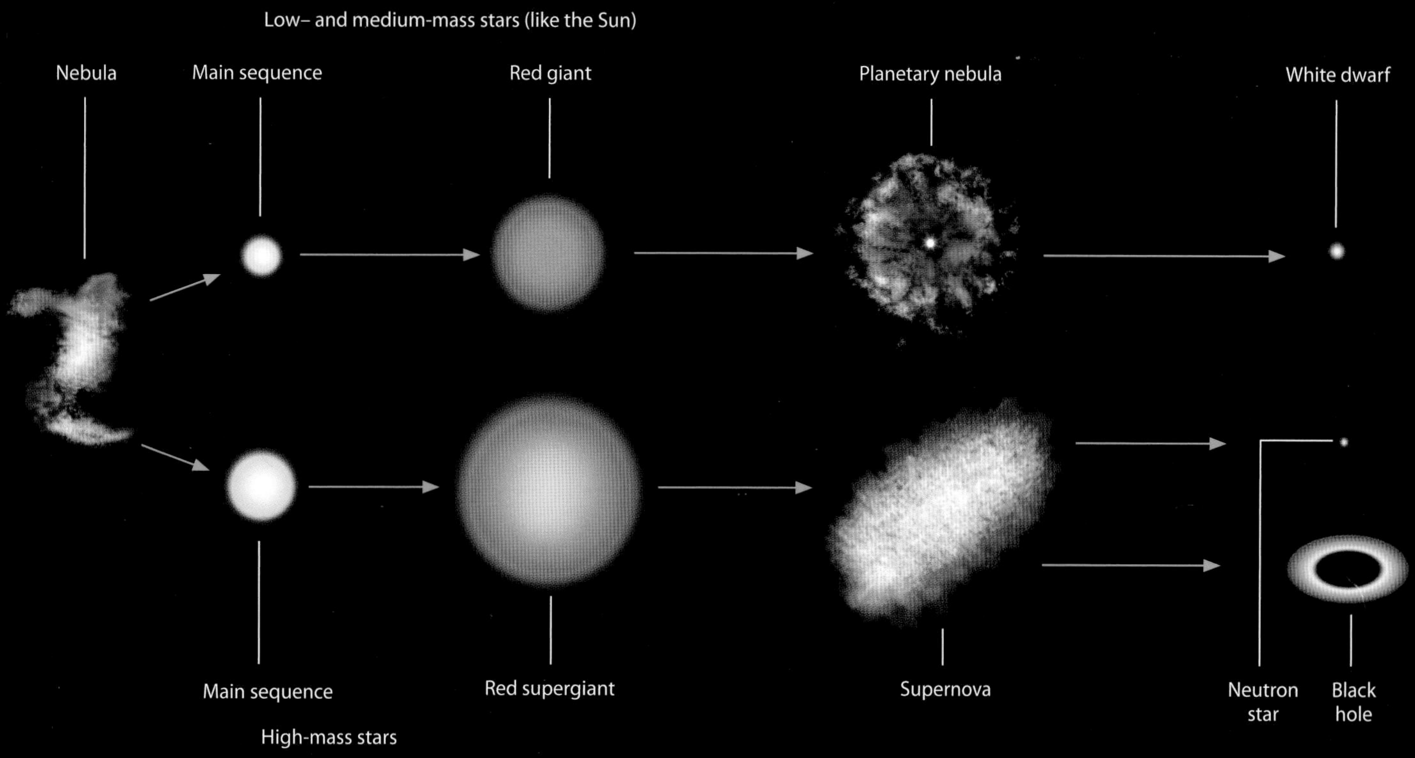

Low– and medium-mass stars (like the Sun)

Nebula
Main sequence
Red giant
Planetary nebula
White dwarf

Main sequence
Red supergiant
Supernova
Neutron star
Black hole

High-mass stars

An average star like the Sun will swell up to form a red giant, which is much bigger but also cooler. For a while heavier atoms like helium and carbon will fuse inside, but eventually fusion stops altogether, and the giant's atmosphere just drifts away forming a cloud of material called a planetary nebula. This is full of heavier elements, such as sodium, iron, and neon. At its heart is the white-hot star core, which is called a white dwarf. White dwarfs gradually cool into cold, non-shining and invisible black dwarfs. (There are no black dwarfs yet, they take many billions of years to form and the Universe is too young.)

Stars with more than 1.38 the mass of the Sun go out with a bang, forming a supergiant that implodes under its great weight creating a violent event called a supernova. A smaller supergiant ends up as a neutron star, just a few kilometers across. The star matter has degenerated into an ultradense material made from pure neutrons. The core of the largest stars collapse even more into a black hole. This tiny but immensely heavy object has gravity so strong that nothing, not even light, can escape.

A short history of the Universe

The Universe began with a bang. It was not only big, it happened everywhere, making the whole of space incredibly hot. The history of the Universe is the story of this hot but tiny space, expanding and cooling. As it cooled the Universe evolved into what we see before us today. Cosmologists, the scientists who study the Big Bang and its aftereffects, divide the Universe's history into epochs where certain phenomena dominated.

THE EPOCHS OF THE UNIVERSE

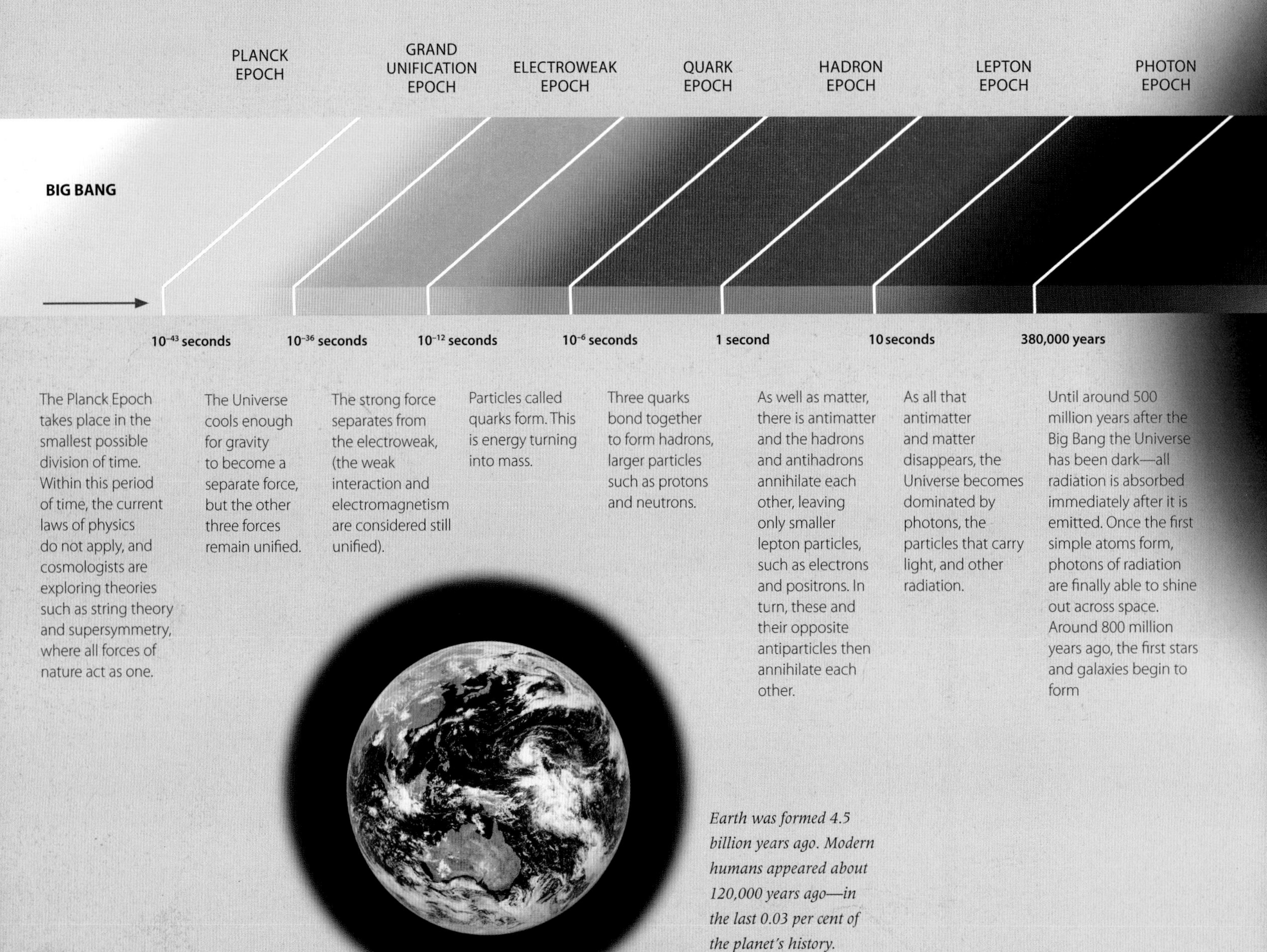

The Planck Epoch takes place in the smallest possible division of time. Within this period of time, the current laws of physics do not apply, and cosmologists are exploring theories such as string theory and supersymmetry, where all forces of nature act as one.

The Universe cools enough for gravity to become a separate force, but the other three forces remain unified.

The strong force separates from the electroweak, (the weak interaction and electromagnetism are considered still unified).

Particles called quarks form. This is energy turning into mass.

Three quarks bond together to form hadrons, larger particles such as protons and neutrons.

As well as matter, there is antimatter and the hadrons and antihadrons annihilate each other, leaving only smaller lepton particles, such as electrons and positrons. In turn, these and their opposite antiparticles then annihilate each other.

As all that antimatter and matter disappears, the Universe becomes dominated by photons, the particles that carry light, and other radiation.

Until around 500 million years after the Big Bang the Universe has been dark—all radiation is absorbed immediately after it is emitted. Once the first simple atoms form, photons of radiation are finally able to shine out across space. Around 800 million years ago, the first stars and galaxies begin to form

Earth was formed 4.5 billion years ago. Modern humans appeared about 120,000 years ago—in the last 0.03 per cent of the planet's history.

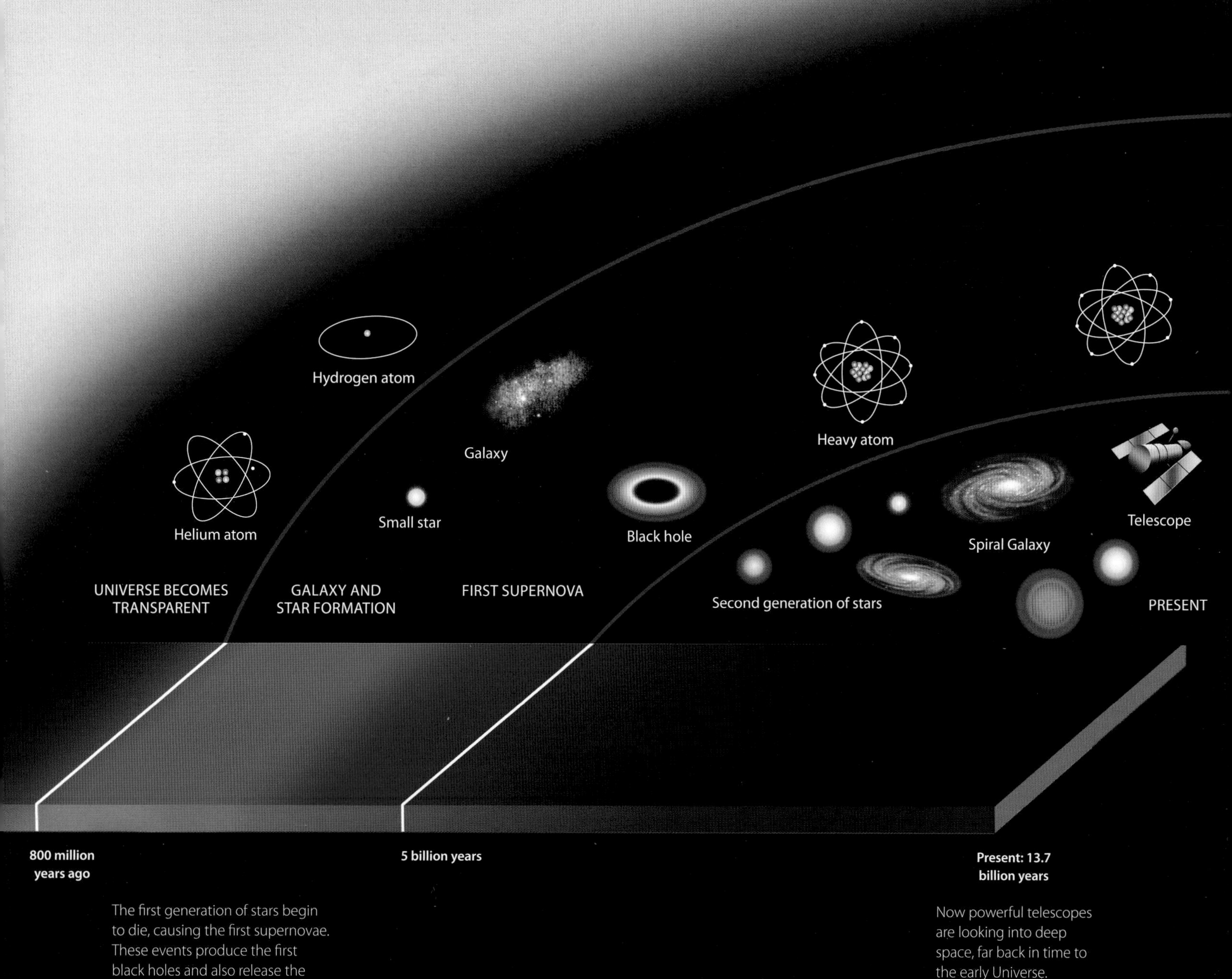

Types of galaxy

There are at least 125 billion galaxies in the Universe (15 each for every person on Earth with some to spare). They come in different shapes and sizes and ages. The youngest are irregular in shape, with many young stars lighting up inside them. As galaxies get older and heavier they form into rotating spirals, such as the Milky Way. Most spirals have central barred sections, where most of the star formation is concentrated. (Some might be amused to hear that the Milky Way has a bar.) Older and larger galaxies, which may have absorbed many small ones, are elliptical.

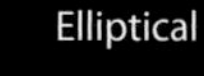

Elliptical

Spiral

Barred spiral

Irregular

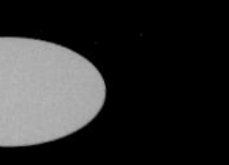

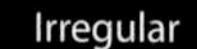

IMPONDERABLES

Perhaps more than any other major science, astronomy is currently in a fascinating time. Recent discoveries have confounded our fundamental ideas about the universe. There is no shortage of mysteries and perhaps we are on the verge of some big discoveries. Whatever answers astronomers find in the future, here are some of the questions.

Do aliens visit us?

Carl Sagan, the great science communicator of the late 20th century, had a phlegmatic response to this one: given the great age of the Universe and given its great size, if it were possible for a hypothetical technology to harness some as yet unknown facet of space and time to travel the great interstellar distances and pay a visit to little old Earth, then the chances are that an ancient and advanced alien culture would have done it by now. He suggested that UFO sightings, on the rise since the invention of flying machines, were a convenient cover for Cold-War military tests. Sagan was not saying that aliens do not exist. He just said that they and us can never meet. The distances are too great. However, he did not say it was impossible that extraterrestrials were on the way, just so unlikely that it is not worth considering. The SETI project (Search for ExtraTerrestrial Intelligence) has been filtering radio waves from space for 35 years, looking for artificial patterns. If a message were announced tomorrow, it is likely to have traveled so far and for so long that its sender lived before the human species had even evolved.

UFOs probably have an explanation a little closer to home than outer space.

Is there a Theory of Everything?

The Theory of Everything is a way of describing how the Universe works using a single system. At the moment we use two: quantum mechanics for the first three forces (strong, weak, and electromagnetic), and the theory of relativity for gravity. The best candidate for a unifying theory is supersymmetry, a descendant of string theory which represents subatomic particles not as zero-dimensional points but as lines, or strings, with many "compact" dimensions. Once captured by mathematics, a string's oscillation defines a particle's properties, such as spin or charge. Supersymmetry suggests there is a connection between bosons, the particles that mediate a force (such as the photon which carries electromagnetism) and the fermions—electrons, quarks, and other particles that give matter its mass. The possibility of symmetry between the massless bosons and the massive fermions could be explored if the much-heralded Higgs boson is found in years to come.

How will the Universe end?

The Universe had a beginning so it is likely to have an end, too. It was popularly thought that if the Universe stopped expanding, it would then contract and end in a Big Crunch, a Big Bang in reverse. If the expansion outran gravity then the Universe would gradually drift apart, losing energy and becoming colder and less active. Eventually heat death would occur, which in keeping with other possible futures is called the Big Freeze. The energy and mass of the Universe would just be so spread out and diffuse that nothing would ever happen ever again. However, it appears that the Universe is expanding ever faster. This is attributed to hypothetical "dark energy," a force that pushes objects away from each other. Its power increases as space expands, so the further everything moves apart, the stronger dark energy becomes, powering the expansion even more. Eventually galaxies will break up as stars are pushed apart. Then stars and planets will be shredded by dark energy. Finally even atoms will be ripped apart in the Big Rip that will spread energy over an infinite space. How long have we got? 22 billion years, give or take.

However it ends, it is going to be Big.

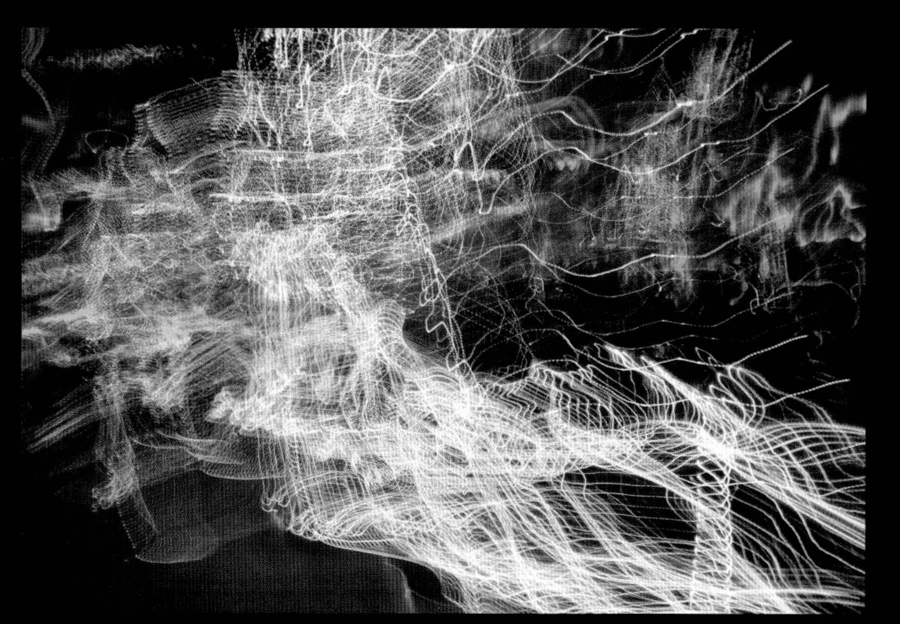

What was before the Big Bang?

The Big Bang started time as well as space so according to that theory there was no "before." One other possibility is that the Big Bang was a Big Bounce. A contracting older Universe collapses to nothingness (a Big Crunch) and then rebounds to do it all over again.

Subatomic particles are not particles (or waves) but strings that vibrate in 10 dimensions, so the theory goes.

IMPONDERABLES

Are there multiverses?

As three-dimensional beings we perceive changes in the dimension above (number four) as constantly changing snapshots of the three dimensions we can see. We call this the passage of time. Imagine if as well as length (the length of time), time also had a width, perhaps better described as splits, bifurcations, alternative paths. Only able to perceive the fourth dimension as passing time, we are oblivious to this fifth dimension, these alternative presents, pasts, and futures that branch off from our single timeline. If every event has more than one outcome—two or more splits in the time line—then it does not take long to end up with a very large number of alternate universes. They have the same laws of physics, it is just that the mass and energy within them occupies a different quantum state. Are these just possible realities or do they actually exist alongside our universe? And can information travel from one universe to another? If so, maybe we'll find out the answer one day.

An alternative view is that daughter universes bubble out of a parent once the parent has expanded to its fullest extent. According to some theories the low energy density produced by a Big Rip resembles the conditions required for a Big Bang.

Do gravitons exist?

The set of subatomic particles that make up the universe comprises what is the Standard Model. One family of particles is called the bosons. Their job is to carry energy between other particles, an event which we measure as a force acting between masses, pushing or pulling them. The boson for electromagnetism is the photon, the gluon mediates the strong interaction, while bosons called W and Z are behind the weak force. It has been suggested that graviton would be a good name for the boson that applies gravity. Gravity is outside the Standard Model—we assume it fits somewhere but don't know how. The Universe might be flooded with gravitons and if we could find one we might discover how the force works. However, gravity is very weak compared to the other forces, and so detecting the effects of these puny bosons has so far proved impossible.

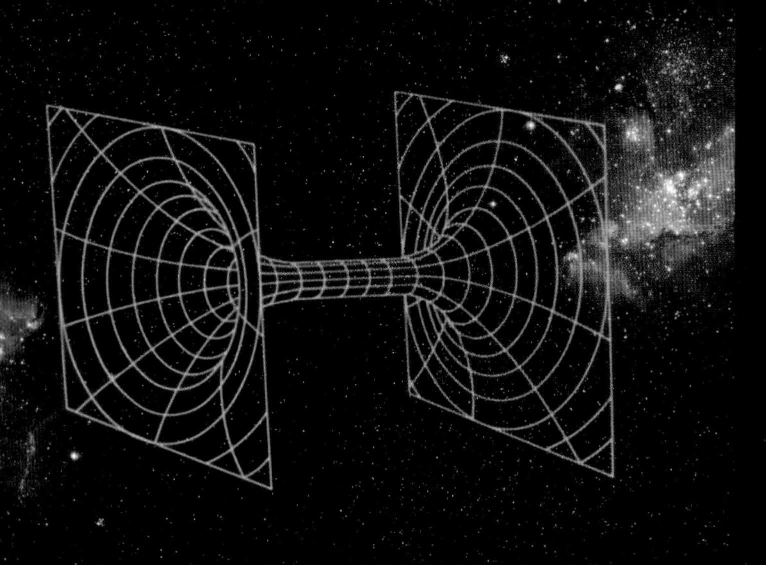

A wormhole might be a black hole with an opening at the other end.

Can we travel to the stars?

The nearest star is a little more than four light-years away. Our rockets can achieve around 4,000th of the speed of light currently so it would be a long trip—getting there alone would be the length of recorded human history. (We should have set off sooner.) There are concepts for faster, longer-lasting engines that could accelerate a starship to much greater speeds, but they only cut the journey to centuries rather than millennia. And the nearest star is a dull red dwarf. Reaching anything interesting would take hundreds, thousands—even—millions of times longer.

However, theoretically it might be possible to travel without moving. Spacetime is warped by mass. If we could somehow reshape the fabric of the Universe into a wormhole, with our destination on one side and our starting point at the other, then we may get somewhere. However, the gravity in such a space would be so strong that your feet would be pulled right through before your head had passed passport control. Could get messy.

Where are all the neutrinos?

As solar fusion squeezes energy out of atoms it also releases a small, shy particle called a neutrino. Our Sun, like every star, is pumping out multiple billions of them every second. The Universe must be full of neutrinos by now. But we have to go to extreme lengths to find even one. The problem is that neutrinos do very little—the term is "weakly interacting." They pass straight through Earth. They are probably streaming through you right now. However, physicists can detect them: all they have to do is fill a mine shaft with radioactive heavy water and wait. When a neutrino arrives it more than likely passes straight through, but very occasionally one hits a water molecule causing a tiny flash. Detectors famously picked up neutrinos emitted by a supernova in 1987. The dying star threw out 10^{58} of these particles. Our scientists picked up 24. The search continues.

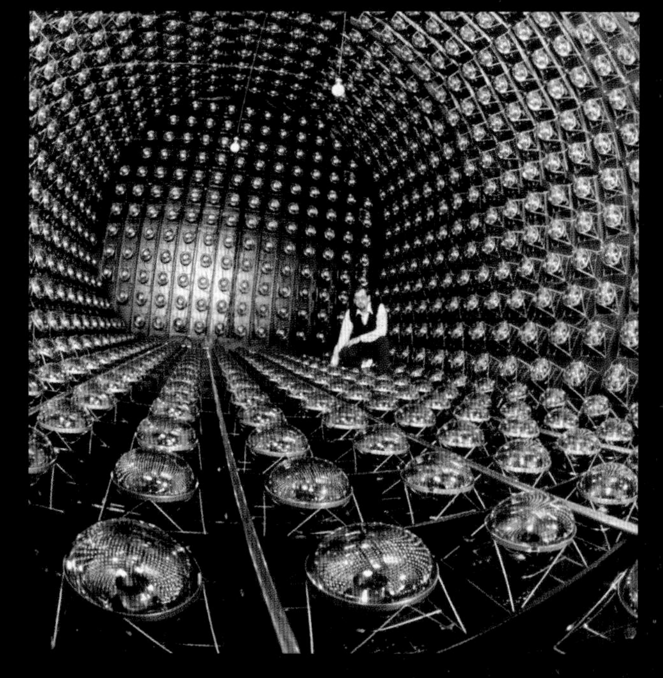

This neutrino detector is lined with sensors that wait for the flash of a passing particle. The detectors are placed underground to shield them from interference from cosmic rays.

The Great Astronomers

The first job of an astronomer is to observe, perhaps with a telescope, perhaps with a radio antenna, or simply with the naked eye—but the greatest astronomers do more. They have been able to interpret those observations, either extracting some universal knowledge from their distant source or revealing the existence of hidden objects. So what makes a person able to do this? Let's take a look at the lives of a few of the great astronomers.

Aristotle

Born	384 BC
Birthplace	Stagira, Greece
Died	322 BC
Importance	Most influential figure in early Western science

The son of the king's doctor, Aristotle was born into Macedonian aristocracy. As befitting his status, he finished his education in Athens, as a pupil of Plato. Aristotle's legacy superseded that of his master and of any other Greek philosopher. It is easy to regard him as a hindrance to scientific enlightenment, since he got a great deal wrong. Nevertheless, Aristotle left us works on poetry, logic, metaphysics, language, and biology that gave intellectuals from as far afield as Turkmenistan and Ireland pause for thought for the best part of two millennia.

Aristarchus

Born	c.310 BC
Birthplace	?
Died	c.230 BC
Importance	First to compare distance to Moon and Sun

Aristarchus of Samos was the first person credited by Nicolaus Copernicus in his Earth-moving account of the heliocentric Solar System. Copernicus must have been inspired by the Greek's work on the relative distances of the Moon and Sun. If not we are in the dark as to what the sixteenth-century Pole was referring to, because this is one of the few proofs of Aristarchus' existence. He was largely ignored by most of his near contemporaries. Archimedes, who was 30 years his younger, was one of the few to mention his astronomical work. However we think that Aristarchus was a student of Strato of Lampsacus, while living in Alexandria—Strato was tutoring the royal princes there, before taking over Aristotle's Lyceum in Athens.

Bell Burnell, Jocelyn

Born	July 15, 1943
Birthplace	Belfast, Northern Ireland
Died	–
Importance	Codiscoverer of pulsars

Bell Burnell, just Bell back then, made her pulsar discovery while still studying for a doctorate at the University of Cambridge. She worked closely with Antony Hewish in setting up a radio telescope and it was her data analysis that led to the breakthrough. But, when Hewish was awarded the Nobel Prize in 1974, Bell Burnell was not so honored (as had been normal for students in the past). However, she has been much lauded since then and continued an illustrious academic career across the world. In 2007 she was elevated to Dame Jocelyn by the British queen.

Bessel, Friedrich

Born	July 22, 1784
Birthplace	Minden, Brandenburg (now in Germany)
Died	March 17, 1846
Importance	Measured distance to stars using parallax

Bessel's career began as an apprentice in the accounts department of a shipping company. He soon became invaluable for his ability to calculate routes for cargos and when he turned this skill skyward and applied it to the motion of Halley's Comet, he won a lowly position at an observatory near Bremen, still only 16. Ten years later he was made director of the Prussian royal observatory in Königsberg. It was at this Baltic city many years later that Bessel became the first, against stiff competition, to measure stellar parallax and so position the stars many millions of kilometers from Earth.

Bethe, Hans

Born	July 2, 1906
Birthplace	Strassburg, Germany (now Strasbourg, France)
Died	March 6, 2005
Importance	Explained solar fusion process

Hans Bethe's Jewish roots forced him to flee Germany in 1933. After a couple of years at English colleges he moved to Cornell in New York. It was here that he and colleagues made their contributions to the understanding of solar fusion in 1939. During the war to follow, Bethe was one of many great minds put to work to harness fission. After the Korean War brought the world to the brink of nuclear conflict yet again, Bethe led the project to build a decisive thermonuclear weapon, or H-bomb, which harnessed the power of fusion to create the biggest explosions in history.

Brahe, Tycho

Born	December 14, 1546
Birthplace	Knudstrup, Scania, Denmark
Died	October 24, 1601
Importance	Last great pre-telescope astronomer

Tycho, as he is often known, enjoyed incredible inherited wealth; at one point his fortune was estimated as one per cent of all Denmark's wealth. While at university Tycho's nose was injured in a duel over the validity of a mathematics formula. He wore a golden implant to fill the hole for the rest of his days. Tycho claimed to have had a pet moose. When asked to present the animal, Tycho said it had died after falling down the stairs drunk. Tycho died from a kidney complaint reportedly brought on when he was too polite to go to the bathroom during a royal banquet in Prague.

Brown, Mike

Born	June 5, 1965
Birthplace	Huntsville, Alabama, U.S.A.
Died	–
Importance	Dwarf-planet hunter

Perhaps not a name that rings through the astronomy hall of fame yet, but Mike Brown has led the charge to find trans-Neptune objects and has found more than any other. A TNO is anything that orbits beyond the last planet and includes Pluto, but also takes in objects that exist beyond the Kuiper Belt. Brown's team has clocked up 14 TNOs in the last 10 years, including Eris, the largest dwarf planet, and Sedna, thought to be the first object seen in the Oort Cloud. Brown's work led directly to the reclassification of Pluto in 2006.

Buffon, Comte de

Born	September 7, 1707
Birthplace	Montbard, France
Died	April 16, 1788
Importance	First scientific estimate of the age of the Earth

Georges-Louis Leclerc, Comte de Buffon, was something of a polymath. He was pondering the ideas of speciation and evolution a century before Darwin, and introduced calculus into probability theory, all while overseeing the French king's botanical garden. He was born without title, and inherited a fortune from his childless godfather. After a roguish youth touring Europe, de Buffon returned to France, claimed his title, and set himself up as a gentlemen scientist in Paris.

Chandrasekar, Subrahmanyan

Born	October 19, 1910
Birthplace	Lahore, India (now in Pakistan)
Died	August 21, 1995
Importance	Calculated minimum mass of supernova

The name Chandrasekhar is forever linked with the Chandrasekhar limit, the minimum mass a star must be to form a supernova. Ironically, his name is derived from the Sanskrit for the "holder of the Moon." Chandrasekhar spent his 20s studying in India before being awarded a postgraduate scholarship at Cambridge. It was on the voyage to this position that he made his famous breakthrough. Chandrasekhar won the Nobel Prize for Physics in 1984 and the orbiting Chandra X-Ray Telescope is named for him.

Cassini, Jean-Dominique

Born	June 8, 1625
Birthplace	Perinaldo, Republic of Genoa (Italy)
Died	September 14, 1712
Importance	First director of the Paris Observatory

Hailing from Genova, in what would later become northern Italy, Cassini is also frequently known as Giovanni Domenico Cassini. Not unusually for the scientists of his age, Cassini was a fan of the occult, and was a noted astrologer as well as being a professor in Bologna. In 1669, in an early example of brain drain, Cassini was lured to Paris by Louis XIV, the self-styled Sun King and not a man to say no to, to head up the new state observatory in Paris. It was while here that Cassini made his mark with observations of Saturn. A gap in the planet's rings is today called the Cassini Division.

Copernicus, Nicolaus

Born	February 19, 1473
Birthplace	Toruń, Poland
Died	May 24, 1543
Importance	Proposed heliocentric model of the Solar System

Copernicus, a Westernized version of Mikołaj Kopernik, was born to a family of merchants, but after the death of his father, the family was taken into the care of his uncle, his mother's brother, a powerful bishop. As well as being a trained doctor and lawyer, Copernicus was fluent in four languages, and followed his elder brother and sister into the clergy, through a post secured by Uncle Lukas. Uncle Lukas also gave his nephew access to many intellectuals of the day, and it is perhaps telling that Copernicus did not start openly discussing heliocentrism until his uncle had died.

Eddington, Arthur

Born	December 28, 1882
Birthplace	Kendal, Westmorland (now Cumbria), England
Died	November 22, 1944
Importance	Proposed theory of stellar fusion

This astrophysicist played a leading role in understanding what stars were made of and how they gave out light. He also helped show that Einstein's theory of relativity was true (despite now disputed results). In 1919 physicist Ludwik Silberstein congratulated Eddington on being one of three men who actually understood the theory of relativity. Silberstein included himself and of course Einstein as the other two. Eddington paused, and when Silberstein encouraged him to agree, the Englishman explained his silence: "I was just wondering who the third one might be!"

Einstein, Albert

Born	March 14, 1879
Birthplace	Ulm, Württemberg, Germany
Died	April 18, 1955
Importance	Formulated theory of relativity

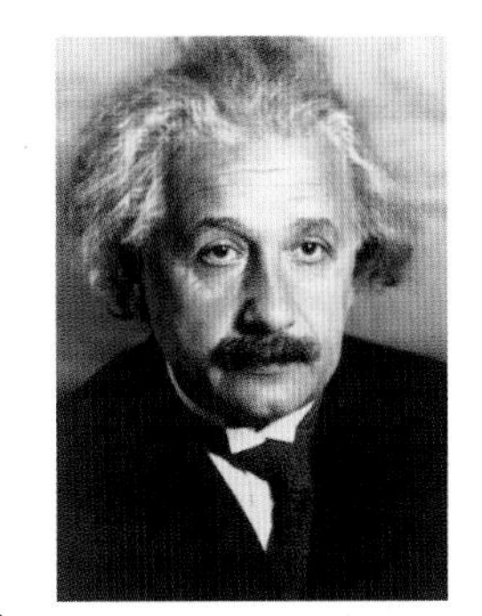

It is often said that Einstein was not highly regarded by his teachers. That was because from an early age he was pursuing his own intellectual agenda. While still a teenager Albert was left to complete his studies in Munich while his parents sought work in Italy—he was not the most attentive pupil. The inconsistent academic record plagued his early career despite his obvious talent. Einstein, now married, took a job as a patent clerk in Bern, Switzerland, in 1903. The untroubling job gave him time to formulate his theories, which two years later propelled him to the top of physics.

Eratosthenes

Born	c.276 BC
Birthplace	Cyrene, Libya
Died	c.194 BC
Importance	Calculated the size of Earth

As the chief librarian at the Great Library of Alexandria, Eratosthenes had the largest information resource the world had ever seen at his fingertips, and he put it to use during his famous measurement of the globe. This feat among others earned Eratosthenes the title of founder of geography, a term he himself coined. The scientist was also a noted early egalitarian, criticising Aristotle's remarks that Greek blood should be kept pure by avoiding marriage with the "barbarian" peoples. Being of North African origin himself, Eratosthenes would probably not have passed Aristotle's heritage test.

Flamsteed, John

Born	August 19, 1646
Birthplace	Denby, near Derby, England
Died	December 31, 1719
Importance	First Astronomer Royal

At the age of 19, Flamsteed wrote a paper on the design and use of astronomical quadrants. Ten years later, having opted for a life in the clergy—a frequent choice for the amateur astronomer of the day—Flamsteed was offered a job as the "King's Astronomical Observator"—better known as Astronomer Royal—at the new Greenwich Observatory. Flamsteed devoted his career to updating the stellar catalog—tripling the number of stars shown by Tycho. In 1712, as the chart neared completion, Isaac Newton and Edmond Halley stole much of it and printed a pirated edition.

Foucault, Léon

Born	September 18, 1819
Birthplace	Paris, France
Died	February 11, 1868
Importance	Experimental physicist, proved rotation of Earth

Best known for his famous pendulum, which now graces museums and science centers across the globe, Foucault also upgraded the apparatus used by his countryman Hippolyte Fizeau for measuring the speed of light. Things could have been very different. He was destined to be a doctor until a phobia of blood ended that ambition. He turned to physics, researching the photographic processes that were cutting edge at the time and dabbling in microscopy. The 1850s were his most productive decade, with work on electrodynamics, optics, and gyroscopes.

Galileo

Born	February 15, 1564
Birthplace	Pisa, Italy
Died	January 8, 1642
Importance	Made first observations with telescope

This scientist is well known as an astronomer and physicist and he was among the first to apply math to their investigation. The son of a musician-cum–mathematician, Galileo chose a career in science, but was always on the look out for a business opportunity—his family always had money troubles. The telescope was one such get-rich-quick scheme, earning him various pensions. However, the description of the Universe that he saw through his telescope put him in conflict with the church, and to avoid jail and secure his income, Galileo was forced to recant his theory that the Earth circled the Sun.

Fraunhofer, Joseph von

Born	March 6, 1787
Birthplace	Straubing, Bavaria (now in Germany)
Died	June 7, 1826
Importance	Founder of spectrometry

Born Fraunhofer but ennobled with a von at the end of his life, this Bavarian was orphaned at the age of 11 and was apprenticed as a glassmaker. At 13 the young Joseph was buried alive when the workshop collapsed. The rescue was led by Maximilian, Prince Elector of Bavaria, who befriended the boy, sponsoring his later education. He was trained as a master lens maker, and discovered a way to make ultraclear optical glass, free of color aberrations. This invention underlay Fraunhofer's spectrometer and other optical devices that were to change the face of astronomy.

Gilbert, William

Born	May 24, 1544
Birthplace	Colchester, Essex, England
Died	December 10 (Nov. 30, old style), 1603
Importance	Discovered Earth's magnetic field

Pre-Newton but post-Copernicus, Gilbert's work on magnetism was seized upon as a possible source of the invisible impetus that could be powering the heavens. However, Gilbert had other concerns. A doctor by training, he was appointed as the royal physician to Elizabeth I until her death in 1603, whereupon he was put in charge of the health of her successor James—who was the first king of both England and Scotland. The history of Britain could have then taken a very different turn. Gilbert himself died a few months later from the plague, but did not kill off the new uniting monarch.

Goddard, Robert

Born	October 5, 1882
Birthplace	Worcester, Massachusetts, U.S.A.
Died	August 10, 1945
Importance	Invented liquid-fueled rocket

Goddard is a national hero in the United States, enough to have a NASA space center named for him. But that did not stop his family from suing the government for patent infringement in 1951. Military rocketry developed by former Nazi scientists had many similarities to Goddard's early designs, as reported by Goddard himself when he inspected a captured V-2 in 1945 months before his death. The patent case rumbled on for nearly a decade, in which the late Goddard was showered with praise, such as honorary gold medals as well as the space center. In the end it took $1 million dollars, an enormous pay-out for 1960, to settle the case.

Harrison, John

Born	March 1693
Birthplace	Foulby, Yorkshire, England
Died	March 24, 1776
Importance	Invented accurate clocks for navigation

A carpenter from rural Yorkshire turned master watchmaker to the king, John Harrison had to struggle every step of the way as he rose up the social and scientific ladder of eighteenth-century England. Harrison's life's work was to secure the Longitude Prize—and so make a fortune. But although his marine chronometers were repeatedly shown to be accurate for navigation, he was regularly rebuffed. His clocks had a "going rate," a known gain or loss of time, which had to be included when making longitude calculations. This made them not good enough for the astronomers who administered the prize.

Halley, Edmond

Born	November 8, 1656
Birthplace	Shoreditch, London, England
Died	January 14, 1742
Importance	Proved comets are orbiting bodies

Edmond Halley is best known for predicting the return of the comet named after him—and proving that not just planets went around the Sun. However, he made other contributions to science. From 1676 to 1678 he traveled to the South Atlantic and made the first accurate maps of the southern skies. On the voyage he also made measurements of Earth's magnetic field particularly paying attention to its direction. The chart he compiled from this data was intended to help sailors estimate longitude. In 1720 he became the second British Astronomer Royal.

Hawking, Stephen

Born	January 8, 1942
Birthplace	Oxford, England
Died	–
Importance	Discovered radiation from black holes

Trapped in a wheelchair due to a nerve disease that also robs him of natural speech, Stephen Hawking has become almost as much of a scientific icon as Albert Einstein, world famous as the brainiac who speaks through a computer. Hawking's contribution to astronomy is the 1974 discovery that even black holes radiate energy. Virtual particles of matter and antimatter exist everywhere, ceaselessly forming and then annihilating each other. At the event horizon of a black hole, these pairs become separated in the instant of their formation and one is released from the hole as Hawking radiation.

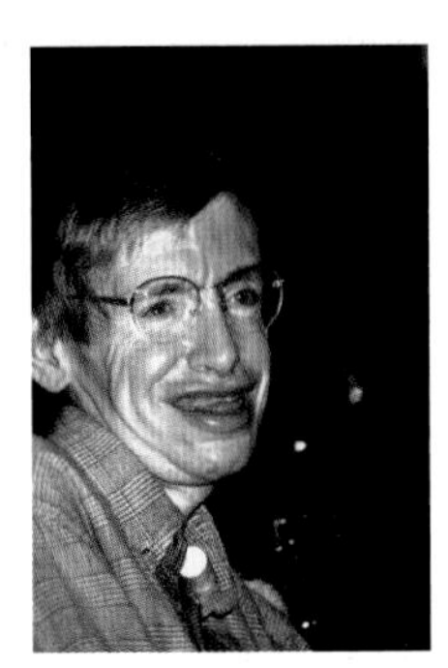

Herschel, William

Born	November 15, 1738
Birthplace	Hannover, Hanover (now Germany)
Died	August 25, 1822
Importance	Discovered Uranus

Wilhelm Herschel's father was the leader of a regimental band in the Hanoverian army. When old enough Wilhelm and his brother Jacob were enlisted in the band; Wilhelm played the oboe. However, the Battle of Hastenback (Hanover lost) turned Wilhelm against a military career, and he went to live in England. He supported himself as a music teacher, before becoming an orchestral leader in Bath. There William, as he was known by now, was joined by his younger sister, Caroline Lucretia, and they formed an astronomical double act that took them from ardent amateurs to professionals in the employ of the king.

Hoyle, Fred

Born	June 24, 1915
Birthplace	Bingley, Yorkshire, England
Died	August 20, 2001
Importance	Codiscoverer of nucleosynthesis

Fred Hoyle was one of the early popularizers of "big picture" astronomy. A natural communicator, he made frequent radio and television appearances as he and his theories rose to prominence. He was well known in his native England for never losing his rural Yorkshire vowels as he spoke on subjects of great enormity. Hoyle is often remembered as an opponent of the Big Bang, preferring a steady-state theory, which he developed with two colleagues he met while building radars during World War II. However, his major contribution was explaining the formation of atoms within stars.

Hipparchus

Born	?
Birthplace	Nicaea, Bithynia (now Iznik, Turkey)
Died	after 127 BC
Importance	Developed trigonometry

Hipparchus developed what would become the field of trigonometry in order to explain the motion he observed in heavenly bodies. Hipparchus spent much of his life on the Aegean island of Rhodes (off the Turkish coast but still part of Greece). His hunch was that the planets moved around the Sun, and he was the first to calculate their motion. However, the results indicated that the planets did not move in perfect circles, causing Hipparchus to abandon the idea as manifestly incorrect: the universe was perfect and so therefore must be its motion.

Hubble, Edwin

Born	November 20, 1889
Birthplace	Marshfield, Missouri, U.S.A.
Died	September 28, 1953
Importance	Discovered that the Universe is expanding

Hubble's interest in science and the stars began in the pages of science fiction, especially the works of Jules Verne, one of the first authors to tell stories of journeys into space. He excelled in athletics as well as math and science during his college career. He won a Rhodes Scholarship to Oxford and turned his back on science, opting for a short but unhappy career as a lawyer. After serving in France during World War I Hubble took a job at the Mount Wilson Observatory, just as the giant Hooker Telescope was being installed, giving Hubble and his colleagues the best view in the world.

Huygens, Christiaan

Born	April 14, 1629
Birthplace	The Hague, Netherlands
Died	July 8, 1695
Importance	Discovered Saturn's rings

One of the great polymaths of the Enlightenment, Huygens is remembered for his work on pendulums and optics, as well as for his astronomical discoveries. He built the first clock to use a swinging pendulum to keep time, and he was the lead advocate of the wave theory of light—in opposition to Newton's corpuscular (particle) theory. (In the end they were both right.) Huygens was also one of the first scientists to address the possibility of extraterrestrial life. He assumed water would be needed and suggested that the spots he saw on Jupiter were oceans, although frozen ones.

Kepler, Johannes

Born	December 27, 1571
Birthplace	Weil der Stadt, Württemberg (now Germany)
Died	November 15, 1630
Importance	Discovered that orbits are elliptical

Kepler was the son of a mercenary, who went off to war when Johannes was five, never to return, presumably killed in action. The boy lived in his grandfather's inn, helping to serve customers. He won a place at the staunchly Protestant college in Tübingen and intended to become a Lutheran minister. Religious wars forced Johannes out of Germany to Prague where his career took off as assistant to the ailing Tycho. Kepler remained a devout person, but his revelation that orbits were not perfect circles led to him being excommunicated by the Pope.

Le Verrier, Urbain

Born	March 11, 1811
Birthplace	Saint-Lô, France
Died	September 23, 1877
Importance	Calculations led to discovery of Neptune

Le Verrier was initially a student of chemistry under the great Joseph Louis Gay-Lussac. However, he soon switched to astronomy and later got a job at the Paris Observatory. Records reflect he was not well liked: "I do not know whether M. Le Verrier is actually the most detestable man in France, but I am quite certain that he is the most detested," was how one colleague put it. His unpopularity would explain why none of the Parisian astronomers would work with him to confirm his discovery of Neptune. He was forced to send his data abroad.

Messier, Charles

Born	June 26, 1730
Birthplace	Badonviller, France
Died	April 12, 1817
Importance	Produced catalog of non-stellar objects

Messier had a comfortable upbringing in rural France, despite the loss of his father at a young age. His elder brother took over his education and when the time came to fend for himself, Messier was able to get work with the chief naval astronomer in Paris. His duties included map-making and assisting with observations. He joined the throng of astronomers looking for the return of Halley's Comet, as predicted 76 years before. He was frustrated by all the false sightings, which led to the eventual catalog of non-stellar objects that bears his name.

Newton, Isaac

Born	December 25, 1642 (January 4, 1643, New Style)
Birthplace	Woolsthorpe, Lincolnshire, England
Died	March 20 (March 31), 1727
Importance	Discoverer of law of gravitation

Over and above his work on optics and calculus, Newton's laws of motion and gravity laid a foundation stone for modern physics—they were enough to map a route to the Moon 300 years later. With a childhood marked by the loss of his father and rejection by his mother, Newton the man was secretive, selfish, and vindictive. The fabled apple story is reputed to have happened while at the family home in Lincolnshire in retreat from the Plague that was sweeping through the cities. Newton guarded discoveries so jealously that it was often decades before they were published.

Rømer, Ole

Born	September 25, 1644
Birthplace	Århus, Jutland, Denmark
Died	September 23, 1710
Importance	First astronomical measurement of speed of light

Rømer's given name was Pederson, but his family changed it to the name of their home island (Rømø) to distinguish themselves from others. While studying in Copenhagen, the young Rømer lived with Rasmus Bartholin, a distinguished scientist of the day who was editing the papers of Tycho Brahe, the great Danish astronomer, prior to their publication. After a spell in France tutoring the royal family and working at the Paris Observatory, Rømer returned home to a varied career as police chief and mathematician to the Danish royal court, as well as professor of astronomy at Copenhagen University.

Ptolemy

Born	c.100
Birthplace	?Egypt
Died	c.170
Importance	Author of *Almagest* star catalog

Claudius Ptolemy was a Roman citizen who wrote in Greek, the language of the intellectual in the Roman era—ironic since Latin was the language of choice for later scholars. Despite his name he was not a ruler of Egypt, but the confusion was an easy one since so many Alexandrian pharaohs had that name, and he is often styled Ptolemy the Wise. Although he spent many years in Alexandria, some authorities refer to him as an Upper Egyptian, which means he hailed from the south of the country—the Egyptian "upper' and "lower" are reversed from those expected on a map.

Sagan, Carl

Born	November 9, 1934
Birthplace	Brooklyn, N.Y., U.S.A.
Died	December 20, 1996
Importance	Space probe designer and science advocate

Carl Sagan was the leading astronomy communicator of his generation. Backed up by a past in academia and working at NASA, Sagan's career transformed into TV personality, popular science author, and campaigner against nuclear weapons. His *Cosmos* series, aired in 1980, turned a whole generation on to the theories of astronomy and cosmology. He then became a driving force behind SETI, the Search for ExtraTerrestrial Intelligence, and was at pains to publicize the outcome of an atomic war—coining the phrase "nuclear winter" to describe the post-conflict conditions.

Schwarzschild, Karl

Born	October 9, 1873
Birthplace	Frankfurt am Main, Germany
Died	May 11, 1916
Importance	Calculated the dimensions of a black hole

Schwarzschild was no ordinary child. At the age of 16 he published a paper on celestial mechanics, the motion of heavenly objects. By the age of 23 he had been awarded a PhD for work on multidimensional geometry. As was fitting for such a talent, after a stint at a Vienna observatory, Schwarzschild became the director of the observatory at Göttingen, a position once held by Carl Gauss. Schwarzschild was on the Russian Front in 1915 when he did the work for which he is remembered today. He also developed an autoimmune disease during World War I which eventually killed him.

Tsiolkovsky, Konstantin

Born	September 5 (Sept. 17, New Style), 1857
Birthplace	Izhevskoye, Russia
Died	September 19, 1935
Importance	Proposeed space travel using rockets

Tsiolkovsky was deaf for most of his life after contracting scarlet fever at around the age of 10. This added to his natural propensity for reclusive behavior and he stayed at home schooling himself, mainly from books in his father's library. He taught math at a school in a small town south-west of Moscow, where he was regarded as something of an eccentric. His work is that of a man who spent time alone just thinking. As well as dreaming of spacecraft, he also conceived of the space elevator, a device that would lift people up to a platform in orbit around Earth.

Tombaugh, Clyde

Born	February 4, 1906
Birthplace	Streator, Illinois, U.S.A.
Died	January 17, 1997
Importance	Discovered Pluto

Tombaugh's farming family could not afford to send him to college, and the young Clyde made his own telescopes, grinding lenses and mirrors to his own specifications. He sent the drawings he made to the Lowell Observatory, and so impressive were they, he was offered a job there. As well as Pluto, Tombaugh discovered several asteroids. During World War II he taught navigation at a naval college. He worked in rocket guidance in the 1950s, before seeing out his career as an astronomy professor at New Mexico State University.

Zwicky, Fritz

Born	February 14, 1898
Birthplace	Varna, Bulgaria
Died	February 8, 1974
Importance	Discovered dark matter and proposed supernovae

Zwicky, half-Swiss, half-Czech, born in Bulgaria, spent most of his life in California. He married the daughter of a wealthy senator, and his wife's money ensured the smooth running of Caltech's observatory in the Palomar Mountains. Zwicky was able to install one of the first Schmidt telescopes there in the 1930s, which was used for a wide-angle survey of the heavens. The telescope aided in the hunt for the first supernovae. Outside of astronomy Zwicky worked on early jet engines and rockets. It is said that one of his experiments launched a metal pellet into the first solar orbit (by accident).

BIBLIOGRAPHY AND OTHER RESOURCES

Books

Atkins, P.W. *Galileo's Finger: The Ten Great Ideas of Science.* Oxford: Oxford University Press, 2004.

Aughton, Peter. *The Story of Astronomy.* London: Quercus, 2008.

Chaisson, Eric J. and Steve McMillan. *Astronomy Today.* San Francisco: Benjamin Cummings, 2010.

Cox, Brian and Andrew Cohen. *Wonders of the Universe.* London: HarperCollins, 2011.

DeVorkin, David H. *Beyond Earth: Mapping the Universe.* Washington, D.C.: National Geographic Society/Smithsonian Institute, 2002.

Dickinson, Terence. *Nightwatch: A Practical Guide to Viewing the Universe.* Richmond Hill: Firefly, 1998.

Genta, Giancarlo and Michael Rycroft. *Space, the Final Frontier?* Cambridge: Cambridge University Press, 2003.

Hawking, Stephen. *A Brief History of Time.* London: Bantam, 1988.

Hoskin, Michael (ed.). *The Cambridge Illustrated History of Astronomy.* Cambridge: Cambridge University Press, 1997.

Lang, Kenneth R. *The Cambridge Guide to the Solar System.* Cambridge: Cambridge University Press, 2003.

MacArdle, Meredith (ed.). *Scientists: Extraordinary People who Changed the World.* London: Basement Press, 2008.

Maran, Stephen P. *Astronomy for Dummies.* Hoboken: Wiley, 2005.

Rees, Martin (ed.). *Universe.* London: Dorling Kindersley, 2008

Suplee, Curt. *Milestones of Science.* Washington, D.C.: National Geographic Society, 2000.

Museums

American Museum of Natural History, New York, U.S.A. www.amnh.org

Armstrong Air & Space Museum, Wapakoneta, Ohio, U.S.A. www.ohsweb.ohiohistory.org

Astronomy Museum, University of Bologna, Italy. www.bo.astro.it

California Science Center, Los Angeles, U.S.A. www.californiasciencecenter.org

Canada Science and Technology Museum, Ottawa, Canada. www.sciencetech.technomuses.ca

China Science and Technology Museum, Beijing, China. www.cstm.org.cn

Cité des Sciences et de l'Industrie, Paris, France. www.cite-sciences.fr

Copernicus Science Centre, Warsaw, Poland. www.kopernik.org.pl/en/

Deutsches Technikmuseum/German Museum of Technology, Berlin, Germany. www.sdtb.de

Herschel Museum of Astronomy, Bath, U.K. www.bath-preservation-trust.org.uk

Hong Kong Space Museum, China. www.lcsd.gov.hk

Huygensmuseum Hofwijck, Netherlands. www.hofwijck.nl

Kansas Cosmosphere and Space Center, Hutchinson, Kansas, U.S.A. www.cosmo.org

Kepler Museum, Prague, Czech Republic. www.keplerovomuzeum.cz

Kepler-Museum Weil der Stadt, Germany. www.kepler-museum.de

Memorial Museum of Astronautics, Moscow, Russia. www.russianmuseums.info

Museo Galileo, Institute and Museum of the History of Science, Florence, Italy. www.museogalileo.it

Museum of Flight, Seattle, U.S.A. www.museumofflight.org

National Air and Space Museum, Smithsonian Institution, Washington D.C., U.S.A. www.si.edu

Powerhouse Museum, Sydney, Australia. www.powerhousemuseum.com

San Diego Air and Space Museum, California, U.S.A. www.sandiegoairandspace.org

Science Museum, London, U.K. www.sciencemuseum.org.uk

Shanghai Science and Technology Museum, Shanghai, China. www.sstm.org.cn

Space Expo/European Space Agency visitor centre, Noordwijk, Netherlands. www.spaceexpo.nl

Taipei Astronomical Museum, Taiwan. english.tam.taipei.gov.tw

Tanegashima Space Center, Kagoshima, Japan. www.jaxa.jp

Telus World of Science, Edmonton, Canada. www.edmontonscience.com

Tsiolkovsky State Museum of the History of Cosmonautics, Kaluga, Russia. www.gmik.ru

Tycho Brahe Museum, Island of Ven, Sweden. www.tychobrahe.com

NASA Visitor Centers

Great Lakes Science Center, Cleveland, Ohio, U.S.A. www.glsc.org

INFINITY Science Center, Stennis Space Center, Mississippi, U.S.A. www.infinitysciencecenter.org

Jet Propulsion Laboratory, California Institute of Technology, Pasadena, U.S.A. www.jpl.nasa.gov

Kennedy Space Center Visitor Complex, Florida, U.S.A. www.kennedyspacecenter.com

Space Center Houston, Texas, U.S.A. www.spacecenter.org

U.S. Space & Rocket Center/Space Camp, Huntsville, Alabama, U.S.A. www.spacecamp.com

Virginia Air and Space Center, Hampton, Virginia, U.S.A. www.vasc.org

Observatories and Planetariums

Adler Planetarium, Chicago, U.S.A. www.adlerplanetarium.org

Alpine Astrovillage Lü-Stailas, Switzerland. www.alpineastrovillage.net

Athens Planetarium, Greece. www.eugenfound.edu.gr

Griffith Observatory, Los Angeles, U.S.A. www.griffithobs.org

Lick Observatory, Mt. Hamilton, California, U.S.A. www.mthamilton.ucolick.org

Melbourne Planetarium, Melbourne, Australia. www.museumvictoria.com.au/planetarium

Paranal Observatory, Atacama Desert, Chile. www.eso.org/paranal

Paris Observatory, Paris, France. www.obspm.fr

Prime Meridian, Planetarium, and Royal Observatory, Greenwich, London, U.K. www.rmg.co.uk

Rome Observatory, Rome, Italy. www.mporzio.astro.it

Roque de los Muchachos Observatory, Canary Islands, Spain. www.iac.es

Royal Observatory, Edinburgh, Scotland. www.roe.ac.uk

Subaru Telescope, National Astronomical Observatory of Japan, Mauna Kea, Hawaii. www.naoj.org

Archives and Exhibits

Hans Bethe papers, Cornell University Library, New York, U.S.A. www.rmc.library.cornell.edu

Cassini papers, Bologna University, Italy. www.amshistorica.unibo.it/giovannicassini

Cosmonauts Alley, Moscow, Russia.

Arthur Eddington papers, Trinity College, Cambridge University, U.K. www.trin.cam.ac.uk

Albert Einstein papers, Hebrew University of Jerusalem, Israel www.huji.ac.il

John Flamsteed papers, Cambridge University Library, Cambridge, U.K. www. cudl.lib.cam.ac.uk

Foucault's Pendulum, St. Martin des Champs, Musée des Arts et Métiers, Paris, France. www.arts-et-metiers.net

Fraunhofer telescope, Deutsches Museum, Munich, Germany. www.deutsches-museum.de

Robert Goddard collection, Clark University, Worcester, Massachusetts, U.S.A. www.robertgoddard.clarku.edu

Edmond Halley papers/journals/letters, Cambridge University Library, Cambridge, U.K. www. cudl.lib.cam.ac.uk; British Library, London, U.K. www.bl.uk; Royal Society, London, U.K. www.royalsociety.org

John Harrison drawings, Guildhall Library, London, U.K. www.cityoflondon.gov.uk

Stephen Hawking drafts of *A Brief History of Time*, Cambridge University Library, Cambridge, U.K. www. cudl.lib.cam.ac.uk

William Herschel papers, Royal Astronomical Society, London, U.K. www.ras.org.uk/

Edwin Hubble papers, Huntingdon Library, San Marino, California, U.S.A. www.oac.cdlib.org

Isaac Newton papers, Cambridge University Library, Cambridge, U.K. www. cudl.lib.cam.ac.uk

Ole Rømer pendulum clock, Kroppedal Museum, near Copenhagen, Denmark. www.kroppedal.dk

Emilio Segrè Visual Archives, American Institute of Physics. www.photos.aip.org

Space Shuttle Enterprise, Intrepid Sea, Air, and Space Museum, New York, U.S.A. www.intrepidmuseum.org

Clyde Tombaugh papers, New Mexico State University library, Las Cruces, U.S.A. www.lib.nmsu.edu

Organizations

International Astronomical Union, www.iau.org

Carl Sagan Foundation, www.carlsagan.com

SETI, Search for Extraterrestrial Intelligence, www.seti.org

Websites

NASA: www.nasa.gov

Nobel Foundation: www.nobelprize.org

OpenLearn Virtual Planisphere: http://www.open.edu/openlearn/science-maths-technology/science/physics-and-astronomy/astronomy/virtual-planisphere

Apps

NASA for iPhone, Android

StarMap for iPhone

Star Walk for iPad

INDEX

Cataloging-in-Publication Data has been applied for and may be obtained from the Library of Congress.

ISBN 978-0-9853230-5-9

Series Concept and Direction: Jeanette Limondjian
Design: Bradbury and Williams
Editor: Meredith MacArdle
Picture Research: Caroline Wood
Cover Design: Jokooldesign

Publisher's Note: While every effort has been made to insure that the information herein is complete and accurate, the publishers and authors make no representations or warranties either expressed or implied of any kind with respect to this book to the reader. Neither the authors nor the publisher shall be liable or responsible for any damage, loss or expense of any kind arising out of information contained in this book. The thoughts or opinions expressed in this book represent the personal views of the authors and not necessarily those of the publisher. Further, the publisher takes no responsibility for third party websites or their content.

SHELTER HARBOR PRESS
603 West 115th Street Suite 163
New York, New York 10025

Printed and bound in China by Imago.

10 9 8 7 6 5 4 3 2 1

Science uses measurements of metric meters and kilometers, so in this book astronomical measurements such as the distance between planets are left in kilometers. One kilometer is 0.62 miles.

PICTURE CREDITS
BOOK

©2005 Antikythera Mechanism Research Project Scientific Data of Fragment A of the Antikythera Mechanism 21 bottom. **Alamy** /© Cosmo Condina Mexico 11 top; © Ancient Art & Architecture Collection Ltd 11 bottom; © North Wind Picture Archives 20; © The Art Archive 22; © Pictorial Press Ltd 26 top; © Mary Evans Picture Library 28 left; © The Art Archive 28 right; ©The Art Gallery Collection 35 bottom; © World History Archive 36 bottom; © Greg Balfour Evans 37 top;© The Art Gallery Collection 39 bottom; © The Art Archive 43 top; © Ian M Butterfield (Bristol) 59 bottom ;© Keystone Pictures USA 62;©The Art Archive 101 top; ©The Art Gallery Collection 5; ©The Art Gallery Collection 34-35 bottom. **Bradbury and Williams** 8-9,18, 19, 21, 61, 65, 109, 118-119, 122-123, 124-125. **Corbis** 10; 14; 44 bottom;/Stocktrek Images 12; The Print Collector 13 bottom; Heritage Images 56 bottom; 130 bottom left; 136 bottom left; 138 bottom right. **Getty** Historic Map Works LLC 59 top. **NASA** 110; **NASA/ JPL-Caltech** 96 top; **NASA/JPL-Caltech/ESA/Harvard-Smithsonian CfA** 68-69; **NASA/JPL-Caltech/UCLA** 2-3. **James Reynolds & Sons** (c. 1850) 7 top. **Science Photo Library** endpapers/Frank Zullo 6 left; SOHO/ESA/NASA 7 bottom; Dr Fred Espanak 17; 19; Royal Astronomical Society 23 top; NYPL/Science Source 23 bottom; European Southern Observatory 25; Detlev van Ravenswaay 27; 29 left; 29 right; 30; Crawford Library/Royal Observatory, Edinburgh 31 top left; top center; 32; Royal Astronomical Society 33 bottom; Adam Hart-Davis 34; 35 top; Royal Astronomical Society 36 top ; NYPL/Science Source 37 bottom; American Institute of Physics 38; Royal Astronomical Society 39 center; Sheila Terry 40; Science, Industry and Business Library/ New York Public Library 41; 42; Royal Astronomical Society 43 bottom; 44 top; Sheila Terry 45; 46 bottom ; Maria Platt-Evans 46 top; 47 top; Royal Astronomical Society 47 bottom; 48 top; Science Source 48 bottom; NOAA 50 bottom; Dr Jeremy Burgess 51; David Parker 52; Royal Astronomical Society 53 top, 53 bottom, 54, 55; Detlev van Ravenswaay 56 top, 57 top; Dept. of Physics, Imperial College 57 bottom; Detlev Van Ravenswaay 58 top, 58 bottom; RIA Novosti 60 bottom; Library of Congress 61 top; 63; Mark Garlick 64; Jon Lomberg 66; Julian Baum 67 top; 67 bottom; NASA/ESA/STSCI/H.Ford, JHU 68 center right; Emilio Segre Visual Archives/American Institute of Physics 68 bottom; NASA 70 left, 70 right; Royal Astronomical Society 71 bottom; 72; NASA/ESA/STSCI/J.Morse, U.Colorado 73 top left; Stefan Schiessl 73 bottom right; Claus Lunau 74; Seymour 75; Detlev van Ravenswaay 76,77; Emilio Segre Visual Archives/ American Institute of Physics 80 center right, 80 bottom left; RIA Novosti 82 left; Jack Finch 82 bottom right ; RIA Novosti 83; US Air Force 84 top; NASA 84 bottom; NASA/JPL 85; Lynette Cook 87; NASA 88 left, 88 right, 89 top; Philippe Plailly/ Eurelios 89 bottom; NASA 90, 91 top; NRAO/AUI/NSF 91 bottom; RIA Novosti 92 top; NASA 93 center right; Joe Tucciarone 95 top; NASA 95 bottom right; Christian Darkin 96 bottom; NASA 97; 98; Henning Dalhoff 99 top; NASA 99 bottom; Julian Baum 100; European Space Agency 101 bottom; NASA/ESA/STSCI/P.Challis & R.Kirschner, Harvard 102; JPL/NASA 103 top, 103 bottom; NASA/CXC/STSCI/JPL-CALTECH 104; NASA 105; Julian Baum 106 top; David A. Hardy 106 bottom; Ton Kinsbergen 107; NASA 108 left, 108 right; Mike Agliolo 111; Johns Hopkins University Applied Physics Laboratory 112; Detlev van Ravenswaay 113 top; Mark Garlick 113 bottom; Detlev van Ravenswaay 114 left; NASA/JPL-CALTECH/UNIVERSITY OF ARIZONA/TEXAS A AND M UNIVERSITY 115 top; ESA/NASA 116 top; Walter Myers 116 bottom; Lynette Cook 117; John Sanford 120 top right; European Space Agency 120 top far left; US Geological Survey 120 top second from left; NASA 120 center; Kevin A. Horgan 120 center second from right; NASA/ESA/L. Sromovsky, U. Wisc/ STScl 121 center right; NASA 121 far right; Kevin A. Horgan 124 bottom; Richard Kail 126; Mark Garlick 127 top; Henning Dalhoff 128; Victor de Schwanberg 129 top; Los Alamos National Laboratory 129 bottom; Robin Scagell 130 bottom right; 131 top left; Sheila Terry 131 top right; Los Alamos National Laboratory 131 bottom left; NASA 131 bottom right; New York Public Library Picture Collection 132 top left; 132 top right; American Institute of Physics 132 bottom left; NYPL/ Science Source 132 bottom right; Library of Congress 133 top left; Sheila Terry 133 top right; US Library of Congress 133 bottom left; Science Source 133 bottom right; 134 top left; Sheila Terry 134 top right; Science Source 134 bottom left; 134 bottom right; Library of Congress 135 top left; Sheila Terry 135 top right; Science Source 135 bottom left; James King-Holmes 135 bottom right; 136 top left; A. Barrington Brown 136 top right; Hale Observatories136 bottom right; 137 top left; Royal Astronomical Society 137 top right; Sheila Terry 137 bottom left; 137 bottom right; Dr Jeremy Burgess 138 top left; Science Source 138 top right; NYPL/Science Source 138 bottom left; 139 top left; 139 top right; 139 bottom left; Emilio Segre Visual Archives/American Institute of Physics 139 bottom right; NYPL/ Science Source 4; RIA Novosti 2-3 background, 4-5 background; NASA 92-93, 114-115; RIA Novosti 60 top; Physics Today Collection/American Institute of Physics 86 top; Stefan Schiessl 78-79; NASA/CXC/M.WEISS 119. **Thinkstock** 15 bottom;/ Photos.com 16; 26 bottom, Photos.com 31 bottom; 49; 94; 95 bottom left; 121 left; /Goodshoot 121 top; 127 bottom; /Digital Vision 120-121 center; /Stockbyte 86-87. **Werner Forman Archive/**British Museum, London 6 right; Royal Canonry of Premonstratensiens, Strahov, Prague 13 top; Private Collection 15 top; British Museum, London. **Colin Woodman** 33 top, 50, 71 top, 81.

TIMELINES

Alamy/ ©The Art Gallery Collection. **Corbis**; Corbis/ © Alfredo Dagli Orti; The Art Archive; © Arvind Garg; © Christophe Boisvieux; © Flip Schulke; © Francis Dean; © Gianni Dagli Orti; © Greg Smith; Hemis; Heritage Images; © Hoang Van Danh; ©Jon Arnold; JAI; ©J P Laffont; Sygma; © Kevin Schafer; © Leif Skoogfors; © Leonard de Selva; © Leslie Richard Jacobs; © Lindsay Hebberd; © Mark Rykoff; Michael Ochs; © Michel Setboun; © Pablo Corral Vega; © Paul Souders ; © Third Eye Images; © Tim Graham; © Werner Forman; © Xinhua; Heritage Images; National Geographic Society; Ocean; Sygma; The Print Collector. **Science Photo Library/**Allan Morton/Dennis Milon; Babak Tafreshi,Twan ; CCI Archives; Daniel Sambraus; David Parker; Detlev Van Ravenswaay; Dr Jeremy Burgess; Martin Bond; Mehau Kulyk; NASA; NASA/Johns Hopkins University Applied Physics Laboratory/Carnegie Insiitution of Washington; NIBSC; RIA Novosti; Royal Astronomical Society; Science Source; Sheila Terry; Simon Fraser; Tek Image; Walt Anderson, Visuals Unlimited.

Publisher's note: Every effort has been made to trace copyright holders and seek permission to use illustrative material. The publishers wish to apologize for any inadvertent errors or omissions and would be glad to rectify these in future editions.